군사영어

Military English

김정기, 남기봉, 천종웅 공저

도서출판 진영사

군사영어 Military English

김정기, 남기봉, 천종웅 공저

2008년 3월 02일 초판 인쇄
2008년 3월 05일 초판 발행
2019년 1월 21일 개정판 발행
2025년 8월 25일 개정2판 발행

발행인 박 진 영
발행처 도서출판 진영사
인천광역시 부평구 주부토로 236번지 인천테크노밸리 U1 지식산업센터 B동 1507호
전화 : 032)505-4207
팩스 : 032)505-4206
E-mail : 0183734207@hanmail.net
신고번호 : 제2007-000001호

ISBN 978-89-6541-716-3 93390
값: 18,000원

21세기를 맞이하여 세계는 하루가 다르게 급변하고 있으며 정치, 경제뿐만 아니라 군사 분야에 있어서도 국가간 협력의 중요성이 증대되고 있다. 지구상 여러 곳에서는 아직도 전쟁을 하고 있으며, 다국적군들이 확전을 방지하고 평화유지와 치안활동을 위하여 노력하고 있는 현실이다. 세계 각국은 자국의 이익을 위하여 이러한 분쟁에 참여하고 있으며 미국을 중심으로 이루어지고 상황에 맞추어 군사 영어의 중요성은 더 말할 나위가 없다고 본다.

장차 대한민국 국군을 이끌어나갈 군간부로써 외국군과의 연합작전을 위해 영어로 조직과 편성, 계급체계, 작전명령과 전투근무지원 등의 일반적 활동과 자기 소개 및 브리핑 등 의사소통 수단과 방법을 이해하고, 이를 효과적으로 숙달하여 활용하는 능력을 길러야 한다.

이 책의 구성을 살펴보면 PART I에서는 기초 군사영어(Basic Military Terms)를 수록하였으며 PART Ⅱ에서는 유용한 표현(Useful Expressions) 용어를 수록하였다. 또한 PART Ⅲ에서는 작전명령(Operation Orders)을 수록하였고, 책말미에는 이 책에서 사용한 용어들을 우리말과 영어의 자모순으로 정리하였다. 그리고 군에서 독특하게 사용되는 관용 표현을 예문과 함께 우리말 번역을 곁들이고, 약어의 원어도 풀어서 설명을 첨부하였다.

본 책을 출간함에 있어 시대적 변화를 담아내려고 노력하였으나 여전히 부족한 점이 남아있어 앞으로도 더 정진하고자 한다. 늘 물심양면으로 도와주시는 도서출판 진영사 사장님과 모든 직원분께 고마움을 전하고 싶다.

2025년 8월, 도의관에서

편집자 일동

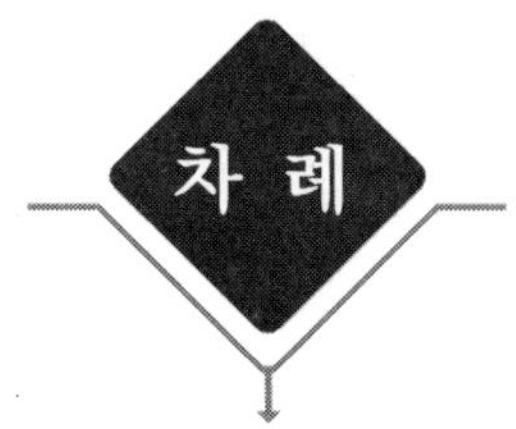

PART I Basic Military Terms

UNIT 1. Military Unit & Symbols ··· 8
UNIT 2. Branches ··· 14
UNIT 3. Ranks ··· 15
UNIT 4. Positions ··· 17
UNIT 5. Weapons ··· 18
UNIT 6. Material and Facilities ··· 20
UNIT 7. Tactics Words ··· 21
UNIT 8. Tactical Control Measures ··· 24
UNIT 9. Offensive Operations ··· 27
UNIT 10. Defensive Operations ··· 28
UNIT 11. Other Operations ··· 30
UNIT 12. Rifle Training ··· 33
UNIT 13. Squad Drill ··· 36
UNIT 14. Communications ··· 38
UNIT 15. Map Reading ··· 40
UNIT 16. War Game ··· 42
UNIT 17. Camp Activity ··· 44
UNIT 18. Duty & Order ··· 46
UNIT 19. Military Service ··· 48
UNIT 20. Unit & Quarter Inspection ··· 50
UNIT 21. Training Program & Test ··· 52

PART Ⅱ Useful Expressions

UNIT 22. Conference Ⅰ 56
UNIT 23. Conference Ⅱ 60
UNIT 24. Office 63
UNIT 25. Telephone Conversation Ⅰ 68
UNIT 26. Telephone Conversation Ⅱ 71
UNIT 27. Restaurant & Party 74
UNIT 28. Personnel 78
UNIT 29. Intelligence 80
UNIT 30. Operations 81
UNIT 31. Operations Plan 82
UNIT 32. Logistics 83
UNIT 33. Engineer 84
UNIT 34. Communications 85
UNIT 35. Policy Planning 86
UNIT 36. Public Affairs 88

PART Ⅲ Examples of Operations Orders

UNIT 37. Company Attack 92
UNIT 38. Company Defense 95
UNIT 39. Armored Division Attack 98
UNIT 40. Infantry Division Defense 103

APPENDICES

A. Glossary (English-Korean) 110
B. Glossary (Korean-English) 137
C. Special Expressions 178
D. Abbreviations and Terms 214

PART I

Basic Military Terms

UNIT 1. Military Unit & Symbols
UNIT 2. Branches
UNIT 3. Ranks
UNIT 4. Positions
UNIT 5. Weapons
UNIT 6. Material and Facilities
UNIT 7. Tactics Words
UNIT 8. Tactical Control Measures
UNIT 9. Offensive Operations
UNIT 10. Defensive Operations
UNIT 11. Other Operations
UNIT 12. Rifle Training
UNIT 13. Squad Drill
UNIT 14. Communications
UNIT 15. Map Reading
UNIT 16. War Game
UNIT 17. Camp Activity
UNIT 18. Duty & Order
UNIT 19. Military Service
UNIT 20. Unit & Quarter Inspection
UNIT 21. Training Program & Test

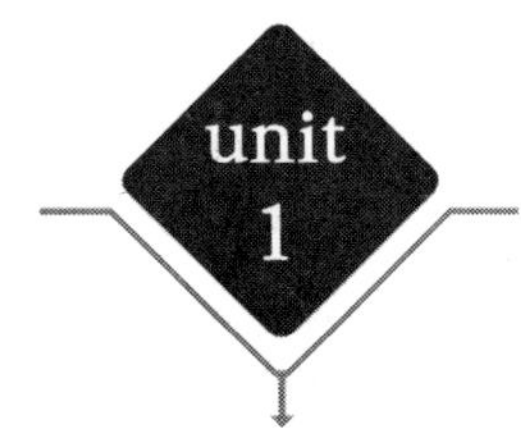

Military Unit & Symbols

단위부대 및 군대부호

■ Units & Military Symbols

Units(단위부대)		Military Symbols(군대부호)
Army	(군) (육군)	XXXX
Corps	(군단)	XXXX
Division	(사단)	XX
Brigade	(여단)	X
Regiment	(연대)	III
Battalion	(대대)	II
Company	(중대)	I
Platoon	(소대)	●●●
Section	(반)	●●
Squad	(분대)	●

■ Basic Symbg1ols(기본부호)

Unit(단위부대)	Observation Post (관측소)
Headquarters(사령부)	Administrative or Logistical Installation (행정 또는 군수시설)
Unit Trains(치중대)	Maintenance Activity (정비활동)
	PW COLL Prisoner of War Collection Pint (전쟁포로 수집소)
Ammunition Supply Point (ASP)(탄약보급소)	Medical Installation (의무시설)

■ Development of a Unit Symbols(부대부호개발)

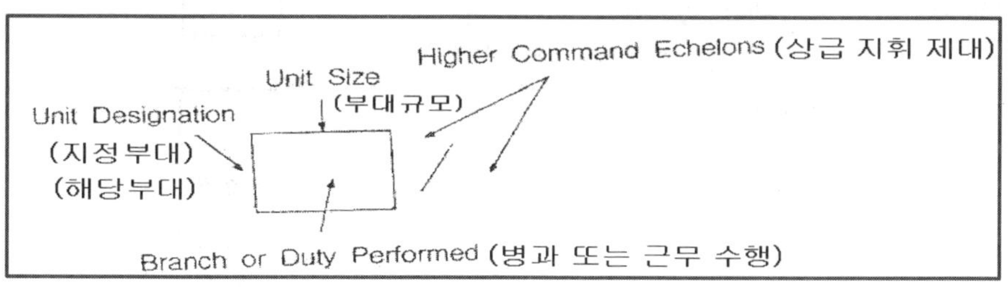

■ Unit Size Symbols(부대규모부호)

● Squad	│ Company, Battery or Troop	Task Force (Bn Size)
● ● Section	Team (Company Size)	│ │ │ Regiment
● ● ● Platoon	│ │ Battalion	╳ Brigade

■ Unit Identifications(부대식별)

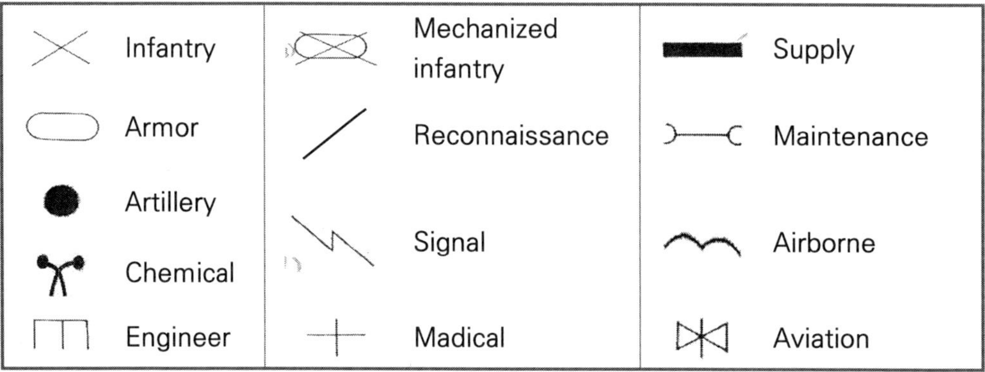

Infantry	Mechanized infantry	Supply
Armor	Reconnaissance	Maintenance
Artillery	Signal	Airborne
Chemical		
Engineer	Madical	Aviation

■ Enemy Units(적부대)

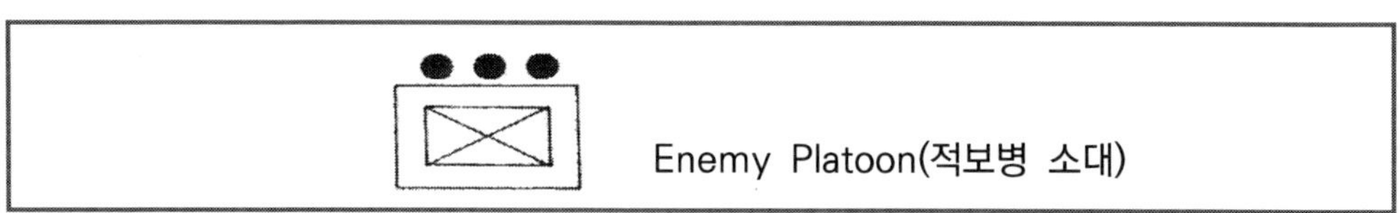

Enemy Platoon(적보병 소대)

■ Future Locations(차후위치, 미래위치)

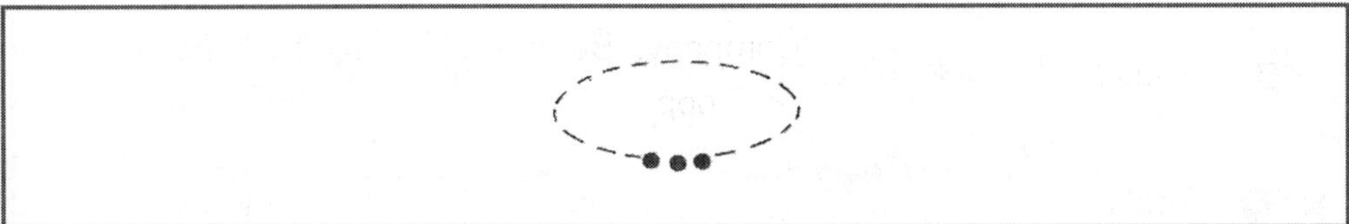

■ Weapons Symbols(화기부호)

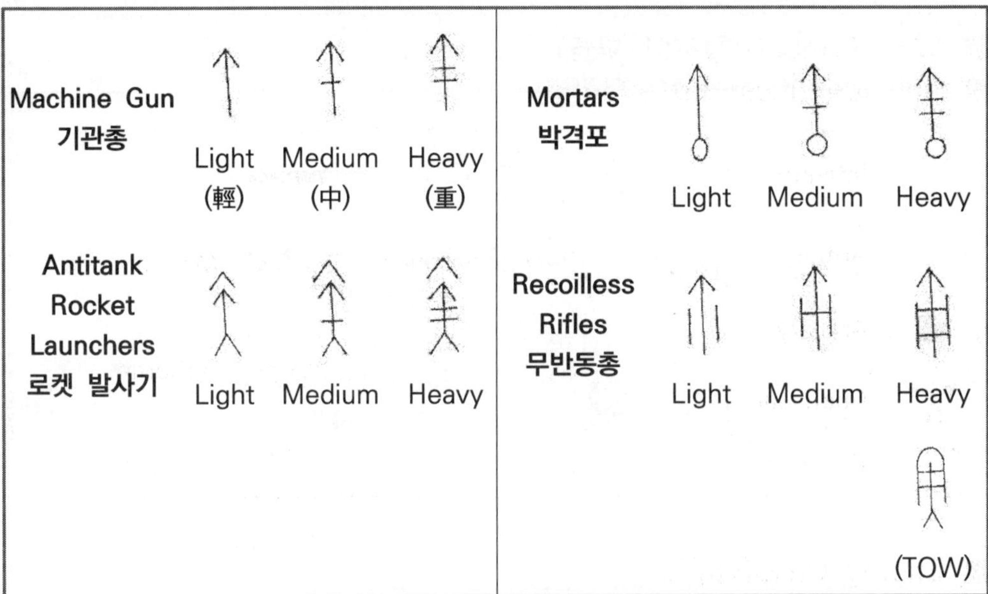

■ Armored Vehicles Symbols(장갑차량 부호)

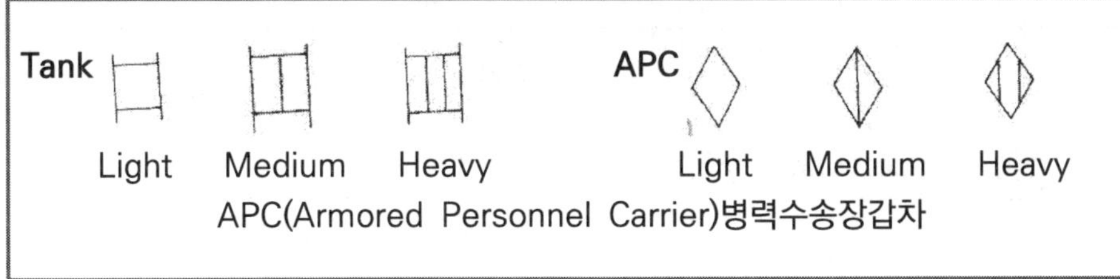

■ Fortifications and Obstacles Symbols(축성 및 장애물 부호)

Foxhole (개인호)

Pillbox (특화점)

Trench (교통호)

Strongpoint (거점)

Tank Ditch (전차호)

■ Wire Symbols(철조망 부호)

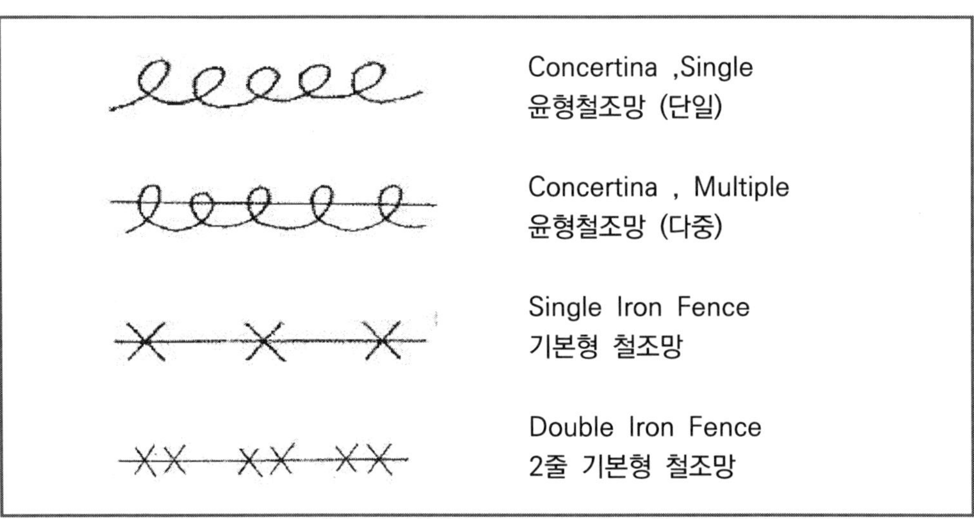

■ Mines Symbols(지뢰 부호)

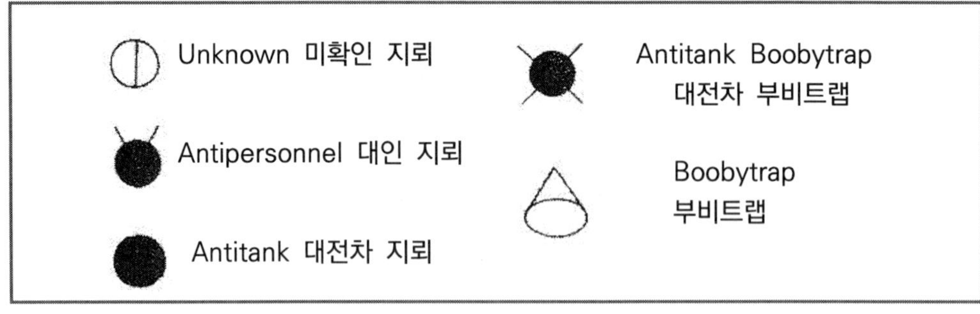

■ Indirect Fire Symbols(간접 사격 부호)

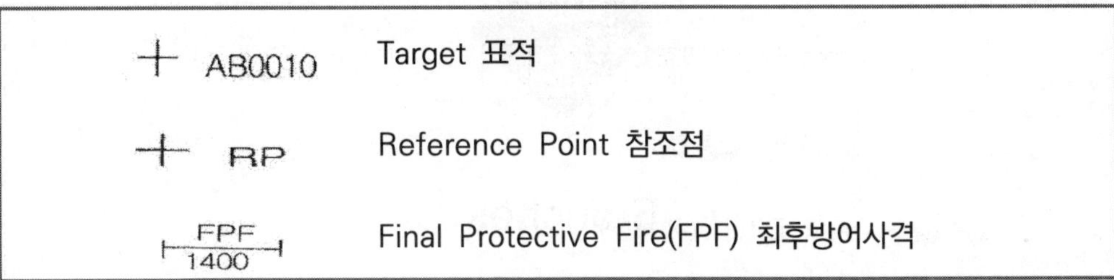

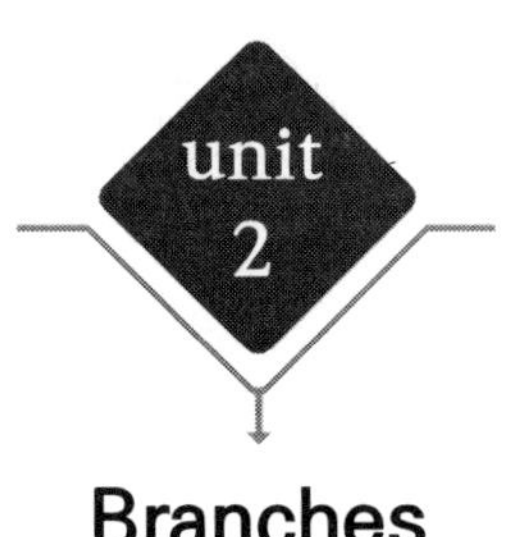

Branches

병과

Infantry (보병)	Field artillery (야전포병)
Armor (기갑)	Air defense artillery (방공포병)
Engineer (공병)	Maintenance (정비)
Communications (통신)	Transportation (수송)
Chemical (화학)	SF Special forces (특전사)
Medical (의무)	
MP Military police (헌병)	
MI Military intelligence (정보)	

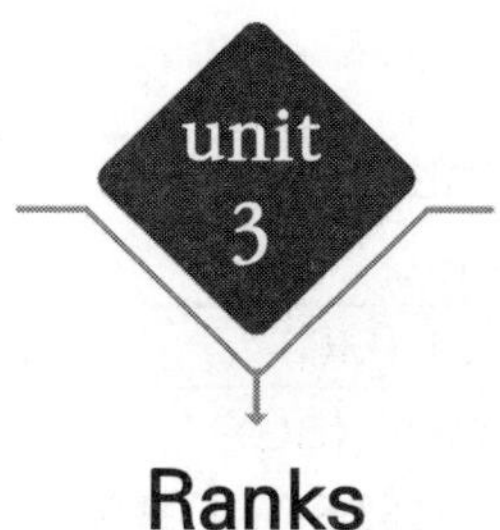

Ranks

계급

■ 병(Enlisted man)

Sergeant (SGT)	병 장
Corporal (CPL)	상 병
Private First Class (PFC)	일 병
Private (PVT)	이 병

■ 부사관(Noncommissioned officer)

Sergeant Major (SGM)	원 사
Master Sergeant (MSG)	상 사
Sergeant First Class (SFC)	중 사
Staff Sergeant (SSG)	하 사

■ 위관장교(Company grade officer)

Captain (CPT)	대 위
First Lieutenant (1LT)	중 위
Second Lieutenant (2LT)	소 위
Warrant Officer (WO)	준 위

■ 영관장교 및 장군(Field grade and general officer)

General (GEN)	대장
Lieutenant General (LG)	중장
Major General (MG)	소장
Brigadier General (BG)	준장
Colonel (COL)	대령
Lieutenant Colonel (LTC)	중령
Major (MAJ)	소령

■ 해군 계급(Ranks for Navy)

Seaman	해군 병사
Petty Officer	해군 부사관
Ensign	해군 소위
Lieutenant	해군 대위
Commander	해군 중령
Caption	해군 대령
Admiral	해군 제독(장군)

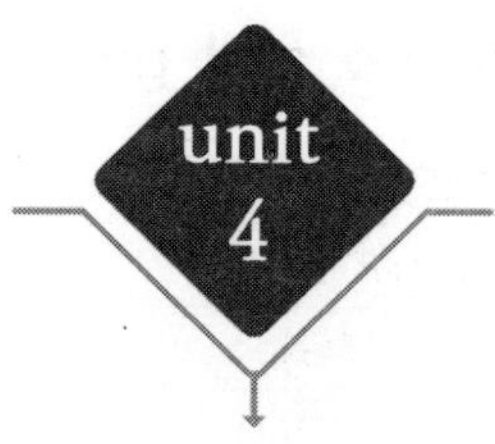

Positions

직책

Chief of Staff	참모총장
Army Commanding General	사 령 관
Corps Commanding General	군 단 장
Division Commanding General	사 단 장
Brigade Commander	여 단 장
Regiment Commander	연 대 장
Battalion Commander	대 대 장
Command Sergeant Major	주임원사
Company Commander	중 대 장
First Sergeant	행정보급관
Platoon Leader	소 대 장
Platoon Sergeant	부소대장
Squad Leader	분 대 장

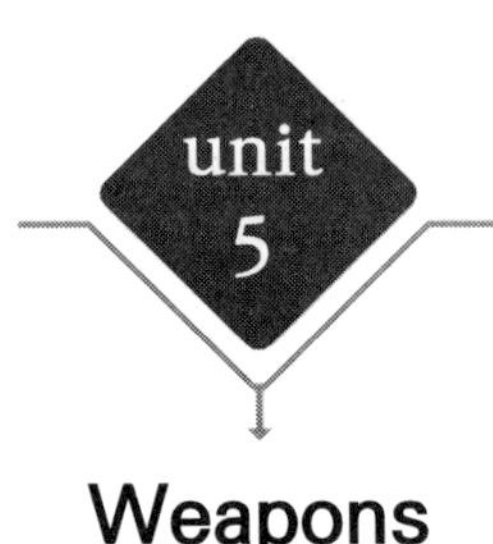

Weapons

무기용어

대검	Bayonet
권총	Hand Gun
소총	Rifle
저격소총	Sniper Rifle
기관단총	Sub Machine Gun
기관총	Machine Gun
수류탄	Hand Grenade
지뢰	Mine
박격포	Mortar
무반동총	Recoilless Rifle
곡사포	Howitzer
다련장 로켓 발사기	Multiple Rocket Launcher
탄도미사일	Ballistic Missile
순항미사일	Cruise Missile
전차	Tank
공격헬기	Attack Helicopter
기동헬기	Airmobile Helicopter
전투기	Combat Aircraft
폭격기	Bomber
수송기	Transport Airplane
항공모함	Aircraft Carrier
잠수함	Submarine
전함	Battle Ship
순양함	Cruiser
생물무기	Bio Weapon

화학무기	Chemical Weapon
대량살상무기	Weapons of Mass Destruction
원자폭탄	Nuclear Weapon

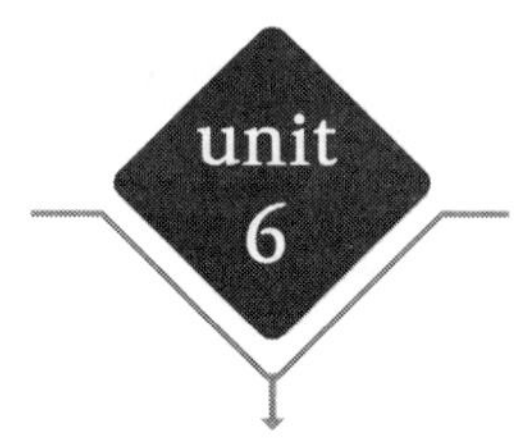

Material and Facilities

물자 및 시설

Class Ⅰ Rations	1종 식량
Class Ⅱ Clothing and Equipment	2종 피복 및 장구류
Class Ⅲ Petroleum, Oil and Lubricants (POL)	3종 유류
Class Ⅳ Construction materials	4종 건축자재
Class Ⅴ Ammunition	5종 탄약
Class Ⅵ Personal demand items	6종 개인물품
Class Ⅶ Major end items	7종 주요 완제품
Class Ⅷ Medical material	8종 의무물자
Class Ⅸ Repair parts and components	9종 수리부속품
Class Ⅹ not included Class Ⅰ through Ⅸ	10종 1종~9종에 포함되지 않는 기타 물자
Base	임시 주둔지
Camp	소규모 기지
Installation	대규모 기지
Airfield	비행장
Guardhouse	위병소
Day room	휴게실
Supply room	보급창고
Crypto center	암호취급소
Drill ground	연병장
Dispensary	의무실
Meals ready to eat(MRE)	전투식량

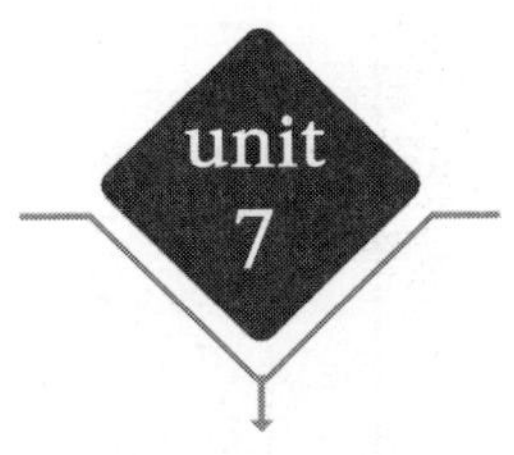

Tactics Words

전술용어

대대전술훈련평가	ATT : Army Training Test
연대전술훈련평가	RCT : Regiment Combat Test
야외기동훈련	FTX : Field Training Exercise
임무형 보호태세	MOPP : Mission Oriented Protective Posture
지휘, 통제, 통신, 컴퓨터, 정보	C4I : Command, Control, Communications, Computer, Intelligence
임무, 적 상황, 지형/기상, 가용병력, 가용시간, 민간요소	METT-TC : Mission, Enemy Information Terrain / weather, Troops&Support available Time available, Civil Consideration
완전군장	Full Combat Gear
피아식별	Identification Friend or Foe
합동작전	Joint Operations
근접항공지원	Close Air Supprt
전쟁준비물자	War Readiness Material
전투근무지원	Combat Service Support
설영대	Quartering Party
증강	Reinforcement
역습	Counter Attack

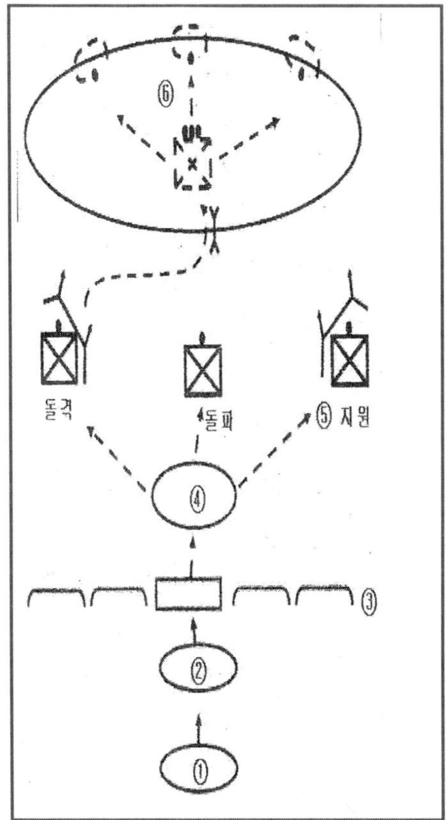

⑥ 목표
: Objective

⑤ 사격지원진지
: Fire support position

④ 돌격진지
: Assault position

③ 공격 개시선/접촉선(LD/LC)
: Line of departure / contact

② 공격대기지점
: Attack position

① 집결지
: Assembly area

Concealment / Cover 은폐 / 엄폐	Battalion Aid Station 대대구호소
Connecting Trench 교통호	Battle Dress Uniform 전투복
Camouflage 위장	ROK/US Combined Forces Command (CFC) 한 · 미연합사
Avenue of Approach 접근로	D-day : Day of attack 공격 개시일
Attack helicopter 공격용헬기	Defense Readiness Condition 방어 준비 태세

Penetration 돌 파	Assault Position 돌 격 진 지
Defensive Position 방 어 진 지	Battle Order 전 투 명 령
Coordination Point 협 조 점	Terrain Intelligence 지 형 정 보
Engagement 교 전	Operation Deception 기 만 작 전
Unit Organization 부 대 편 성	Fire & Maneuver 사격과 기동
Clearing the Field of Fire 사 계 청 소	Final Protective Fire 최후 방어 사격
Dispersion 소 산	Psychological Operation 심 리 작 전
Ammunition 탄 약 할 당	Release Point 분 진 점
General Outpost 일반전초 Guard Post 감시초소	Tactical Operations Center 전술 작전 본부
Anti-Aircraft Artillery 대 공 포	Multiple Launch Rocket 다연장로켓

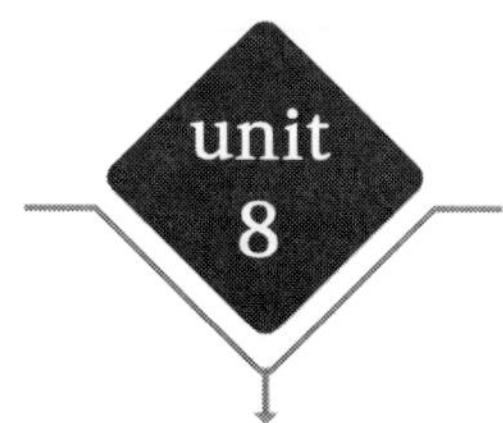

Tactical Control Measures

전술통제수단

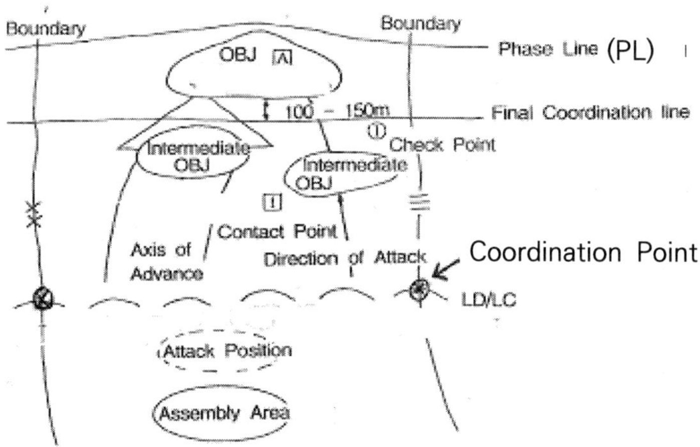

Boundary	전투 지경선
Phase Line (PL)	통제선
Limit of advance (LOA)	전진 한계선
Final Coordination Line	최종협조선
Objective (OBJ)	목 표
Intermediate Objective	중간 목표
Check Point (CP)	확 인 점
Contact Point	접 촉 점
Axis of advance	전 진 축
Direction of Attack	공격 방향
Line of Departure (LD)	공격개시선
Line of Contact (LC)	접촉선
Attack Position	공격대기 지점
Assembly Area (AA)	집 결 지
Coordination Point	협 조 점

Time of attack	공격 개시시간
Maneuver space	기동 공간
Phase	단계
No Fire Line (NFL)	사격 금지선
Fire Support Control Line (FSCL)	사격 지원 통제선
Room for dispersion	산개 공간
Zone of operations	작전지역
Zone of action	전투지역
Avenue of approach	접근로
Area of responsibility	책임지역
Final objective	최종목표
Initial objective	최초목표
Final coordination line	최후협조선(폭격개시선)
No Bomb Line (NBL)	폭격 금지선
Bomb Line (BL)	폭격선
Bomb Safety Line (BSL)	폭격안전선
End of Evening Nautical Twilight (EENT)	해상 박명종
Beginning of Morning Nautical Twilight (BMNT)	해상 박명초
Nuclear Safety Line (NSL)	핵 안전선
Fire Coordination Line (FCL)	사격 협조선

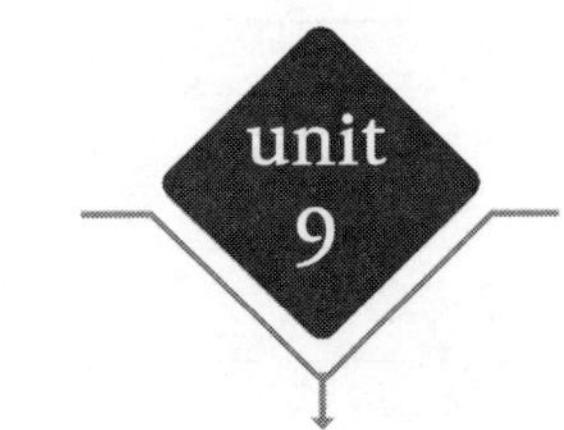

Offensive Operations

공격작전

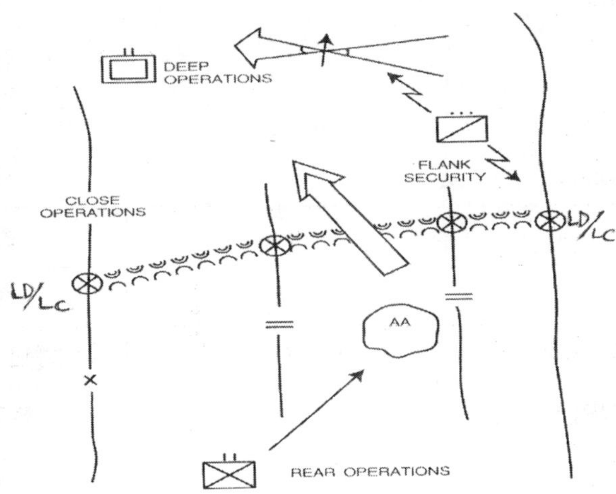

Offensive Operations	공격작전
Deep Operations	종심작전
Close Operations	근접작건
Flank Security	측방방호
Line of Departure (LD)	공격개시선
Line of Contact (LC)	접촉선
Assembly Area (AA)	집결지
Rear Operations	후방 작전
Disposition	배치
Movement to contact	접적 이동
Infiltration	침투
Raid	습격
Scheme of Maneuver	기동계획
Hostile territory	적 지역

Defensive Operations

방어 작전

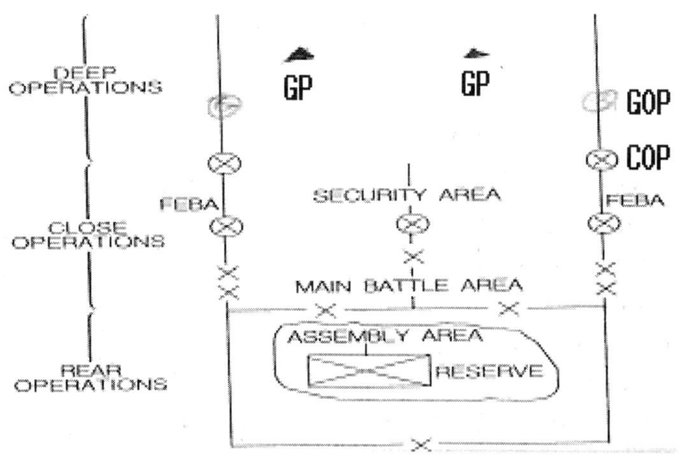

■ DEFCON: Defense Readiness Condition(방어준비태세)

Readiness Condition	Exercise	설명
DEFCON I	Cocked Pistol	최고 수준 전쟁 준비
DEFCON II	Fast Pace	전개 및 투입 준비
DEFCON III	Round House	소집하에 강화된 준비
DEFCON IV	Double Take	감시 강화한 평시
Caption V	Fade Out	평시

■ Security and Defense(경계 및 방어)

Guard Post	경계초소(감시초소)
General Outpost (GOP)	일반 전초
Combat Outpost (COP)	전투 전초
Forward Edge of Battle Area(FEBA)	전투지역 전단
Security Area	경계지역
Main Battle Area	주방어지역
Reserve Area	예비대지역
Assembly Area	집결지
Rehearsal	예행연습
Retrograde operations	후퇴작전
Fields of fire	사계
Likely avenues of approach	예상 접근로
Illumination	조명
Lifting of fire	사격 연신
Shifting of fire	사격 전환
Reverse slope	반사면, 후사면
Grazing fire	최저표척사격
Final Protective Fire	최후방어사격

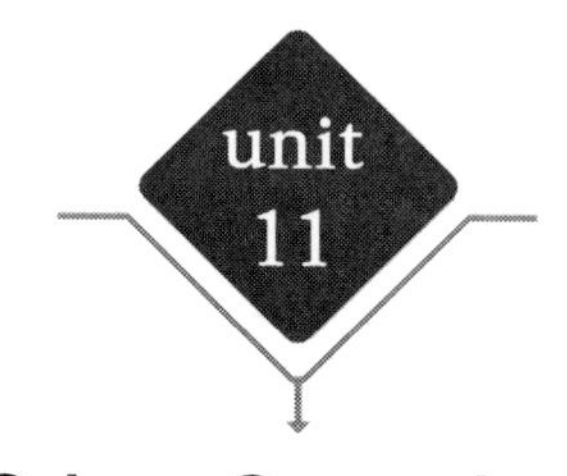

Other Operations

기타 작전

Special Operations	특수작전
Airborne and airmobile operations	공정/공중기동작전
Parachute troops	공수부대
Airmobile tactics	공중 기동전술
Airmobile training	공중 기동훈련
Airborne corridor	공중회랑
Landing Zone (LZ)	착륙지대
Pickup Zone (PZ)	탑재지대
Air defense operations	방공작전
High altitude air defense	고고도 방공
Air-to-air defense	공대공 방어
Antiaircraft defense	대공방어
Air Defense Artillery (ADA)	방공포병
Low altitude air defense	저고도 방공
Medium altitude air defense	중고도 방공
Surface-to-air defense	지대공방어
Unconventional warfare	비정규전
Guerrilla Warfare Operational Area (GWOA)	게릴라 작전지역
Guerrilla warfare	게릴라전
Hit-and-run warfare	유격전
Torture	고문
Operation on interior lines	내선작전
Partisan organization	빨치산 조직
Operation on exterior lines	외선작전

Subversion	전복
Underground resistance	지하저항운동
Riot demonstration	폭동데모
Mountain operations	산악작전
Commanding heights	감제고지
Alpine terrain	고산지대
Rough-terrain suit	산악 침투복
Mountain sickness	산악병
Mountain troops	산악부대
Heavy woods	울창한 수림
Inaccessible terrain	접근불능지역
Limited road net	제한된 도로망
Landing operations	상륙작전
Assault boat	강습 단정
Assault landing team	강습 상륙단
Assault craft	강습 주정
Anti-amphibious minefield	대상륙 지뢰지대
Landing forces	상륙군
Beachhead	해두보(상륙작전시)
Marine Task Force	해병 기동부대
Beach obstacle	해안 장애물
River line operations	하천선 작전
Bridgehead	교두보
Hasty river-crossing	급속 도하
Fordability	도섭능력
River-crossing operations	도하 작전
Deliberate river-crossing	정밀도하
Joint and combined operations	연합 및 합동작전
Military Armistice Commission (MAC)	군사정전위원회
Defense Readiness Condition (DEFCON)	방어준비태세
Ulchi-Focus Lens Exercise	을지포커스 훈련
Armed Forces Korea Tactical Target List(AFKTTL)	주한미군 전술표적 목록
Team Spirit Exercise	팀스피리트 훈련
Military Committee Meeting(MCM)	조사위원회 회의

Security Consultative Meeting (SCM)	한미 안보협의 회의
ROK-US Combined Forces Command	한미 연합군 사령부
Fiont Air Attack Team (JAAT)	합동 공중 공격반
Air alert	공습경보
Orbit point	선회지점
Air Forward Air Controller (A-FAC)	공중 전방항공 통제관
Air Combat Patrol (ACP)	공중 전투초계
Air ground operations net	공지 작전망
Air ground cooperation	공지 협동
Air-Ground Operations System (AGOS)	공지작전 체제
Making panel	대공 포판
Air making	대공표지
Laser Guided Bomb (LGB)	레이저 유도 폭탄
Flight Coordination Center (FCC)	비행 협조 본부
No Fire Line (NFL)	사격 금지선
Tactical air request net	전술 항공 요청망
Tactical Operations Center (TOC)	전술작전본부
Tactical Air Support Element (TASE)	전술항공 지원대
Tactical Air Control Party (TACP)	전술항공 통제반
Tactical Air Control Center (TACC)	전술항공 통제본부
Tactical Air Control System (TACS)	전술항공 통제체제
Ground Controlled Interception (GCI)	지상 관제 요격
Ground alert	지상대기
Ground Forward Air Controller (G-FAC)	지상전방항공 통제반
Joint Air-ground Operation System(JAGOS)	합동공지작전 체제
Joint Operations Center (JOC)	합동작전 본부
How do you hear me?	감도가 어떠한가
I have ~ in sight.	보인다.
No joy.	안 보인다.
Tanks are at your 12 o'clock.	전차는 12시 방향에 있다.
Bring ~ in a little!	좀 더 접근하라.
Target coordinates	표적 좌표

Rifle Training

소총훈련

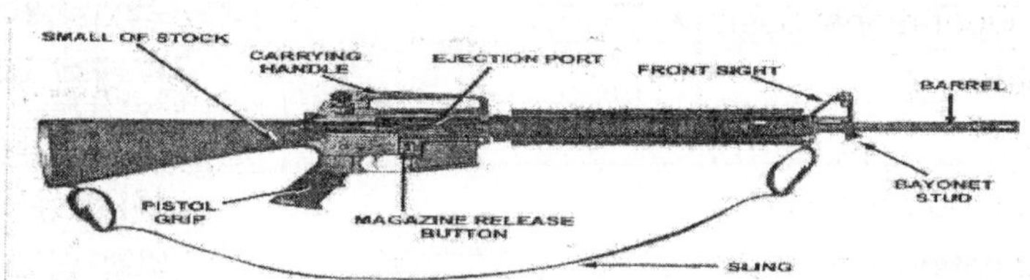

Small of stock	**총목**
Carrying handle	**운반 손잡이**
Ejection port	**방출구**
Front sight	**가늠쇠**
Barrel	**총열**
Bayonet	**대검**
Bayonet stud	**대검 꽂이**
Pistol Grip	**총 손잡이**
Magazine	**탄창(탄알집)**
Magazine Release Button	**탄창해제버튼**
Sling	**멜빵**
Firing range	**소총 사격장**
field range	**야외 사격장**
firing line	**사선(射線)**
firing lane	**사로(射路)**
Control tower	**사격 통제탑**
Pit; maker's gallery	**감적호**
Target	**표적**

Static target; fixed target	고정표적
Static target; moving target	이동표적
Pop-up target	돌연 표적
Bull's eye target	원형흑점표적
Full figure target	전신표적
Ammunition detail	탄약병(彈藥兵)
Target detail	감적수(監的手)
Zero-in	영점조준하다
Kneeling supported position	의탁무릎쏴 자세
Bipod-supported prone position	양각대의탁엎드려쏴 자세
Fire discipline	사격군기
Fire order	사격명령

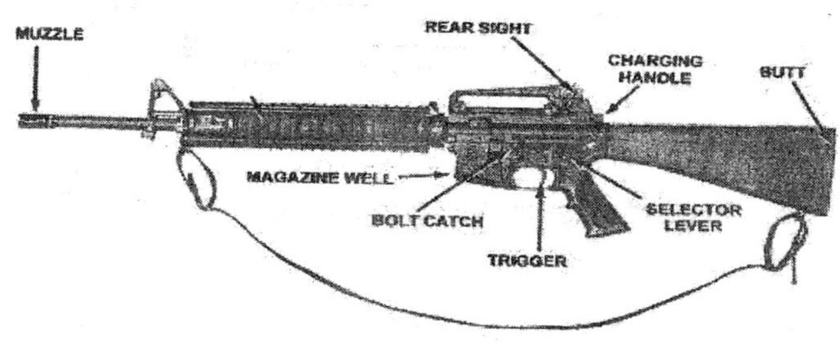

Muzzle	총구
Rear sight	가늠자
Charging handle	장전손잡이
Butt	개머리판
Magazine well	탄창꽂이
Bolt catch	노리쇠 잡이
Trigger	방아쇠
Selector lever	기능선택스위치
Fir command!	사격구령
Close chamber!	노리쇠 전진!
Insert magazine!	탄알집 끼어!
Firers, assume the prone position!	사수, 엎드려 쏴!
Lock, one round!	자물쇠 잠그고 탄알일발장전!
Commence fire!	사격 개시!

Commence firing when ready!	준비된 사수로부터 사격개시!
Cease fire!	사격 그만!
Bore clear!	총구 이상무
Clear on the left!	좌선 사격끝
Firers move down range!	사수 표적 앞으로!
Lock your weapon!	자물쇠 잠궈!
Single shot	단발 사격
Misfire	불발
Accidental discharge	오발
Call the shot	탄착점을 예언하다
Covering fire	엄호사격
Grazing fire	최저 표척사
Oblique fire	사사(斜射)
Plunging fire	감사
Enfilade fire	종사
Flanking fire	측사

Squad Drill

분대훈련

Line formation	횡대대형
Column formation	종대대형
Battalion in line;	대대횡대
Battalion in column;	대대종대
Line of fours	4열 횡대
Column of fours	4열 종대
Rank	오
File	열
Interval	간격
Distance	거리
Alignment	정렬
Parade, rest!	열중 쉬어!
At ease!	쉬어!
Guide on PFC Kim, dress right, dress!	김일병 기준, 우로 나란히!
At close interval, dress right, dress	좁은 간격, 우로 나란히!
(Normal interval,) dress right, dress	정식 간격, 우로 나란히!
At double interval, dress right, dress	양팔 간격, 우로 나란히!
Count off!	번호!
Two step backward, march!	뒤로 2보 갓!
PT formation, move	체조 대형으로 벌려
Stand, fast!	정위치!
Port, arms!	앞에 총!
Present, arms!	받들어 총!
Order, arms!	세워 총!
Right shoulder, arms!	우로 어깨 총!

Sling, arms!	어깨 걸어 총!
Adjust, arms!	멜빵 조여!
Inspection, arms!	검사 총!
Stack, arms!; Plie, arms!	걸어 총!
Take, arms!; Unpile, arms!	풀어 총!

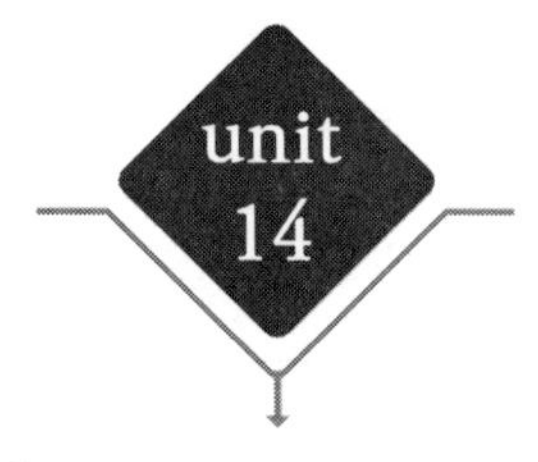

Communication

통신

COMEX (communications exercise)	통신 훈련
Construction center	가설 본부
Wire-laying team	가설반
Signal party	통신 분견대
Signal detachment	통신 파견대
COMMCEN (communications center)	통신 본부, 통신 취급소
Crypto center	암호 취급소
Records center	문서 보관서
Messenger station	전령통신 취급소
Airborne command and communications center	공중 지휘 통신소
Radio man	무전병
Signal man	통신병
Decoder	해역병
Encoder	조립병
Fire direction net	사격지휘망
Headquarters net	본부통신망
Division command net	사단 지휘망
Battalion net	대대 무선망
ACCNET (army command and control network)	육군 지휘 통제망
Signal supported unit	통신 피지원 부대
SOI (signal operation instructions)	통신 운용 지시
SSI (standing signal instructions)	통신 준칙
CEOI (communication-electronics operation instructions)	통신전자운용지시
Communication SOP (standing operating instructions)	통신 내규

RWI (radio-wire integration)	유무선 통합
RWI system	유무선 통합 체제
Wire and radio relay	유무선 중계
Loss of communication	통신 두절

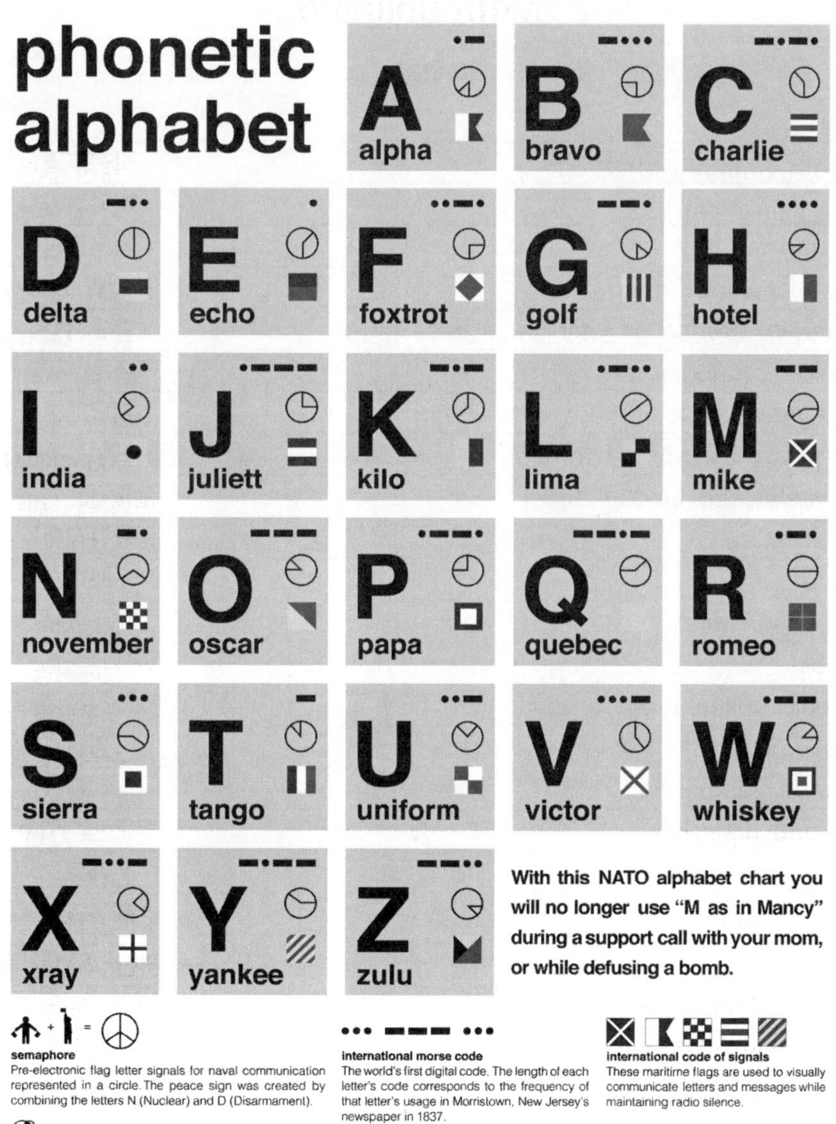

Map Reading

독도법

AO (area of operations)	작전지역
Terrain evaluation	지역평가
Estimate of terrain	지형판단
Enemy occupied area	적점령 지역
Key terrain; ground of vital importance	중요지형
Unfamiliar ground	생소한 지형
Terrain model	지형모형
Five military aspects of terrain	지형평가 5개 요소
Observation and field of fire	관측 및 사계
Dead zone	사각지대
Defilade	차폐
Visual observation	육안 관측
Sky line	공제선
Electronic observation	전차 장비 관측
Dense vegetation	밀립
Cultivated vegetation	식림
Relief and drainage	기복 및 배수
Defiles; bottleneck	애로
Sunken roads	함몰도로
Pocket	소규모 저항지역
Seizing and securing	탈취 및 확보
Avenue of approach	접근로
Compartment of terrain	지형격실
Cross compartment	횡격실
Corridor	종격실

Salient	돌출부
Meteorological station	측후소
WSU (weather support unit)	기상지원부대
SWO (staff weather officer)	기상장교
Weather analysis	기상분석
Climatic information	기후첩보
Weather information	기상첩보
Light data	광명 제원
BMNT (beginning of morning nautical twilight)	해상 박명초
BMCT (beginning of morning civil twilight)	민간 박명초
EENT (end of evening nautical twilight)	해상 박명종
EECT (end of evening civil twilight)	민간 박명종
Visibility	시계
Diurnal range	일교차 (낮시간 범위)
Nocturnal range	야음차 (밤시간 범위)

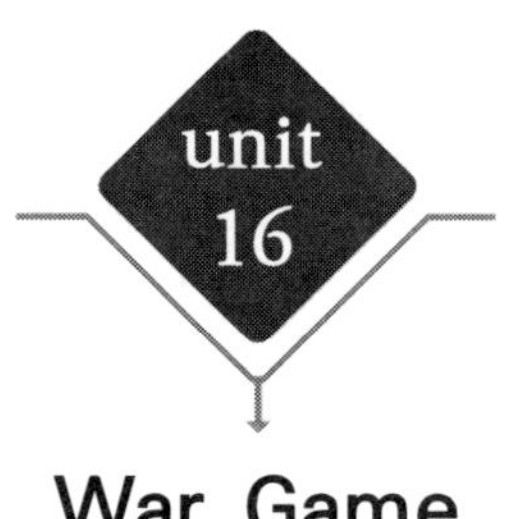

War Game

전쟁연습

Troop-leading procedures	부대지휘절차
Receive mission	임무를 수령하다
Issue a warming order	준비명령을 하달하다
Make a tentative plan	최초계획을 수립하다
Initiate movement	부대이동을 개시하라
Reconnoiter	정찰하다
Complete the plan	계획을 완성하다
Issue the order	명령을 하달하다
Supervise and refine	감독 및 보안하다
Estimate of the situation	상황판단
Detailed mission analysis	세부임무분석
Situation and courses of action	상황 및 방책
Analysis of situation	상황 분석
Development of courses of action	방책발전
Analysis of courses of action	방책분석
Comparison of courses of action	방책비교
Recommendation and decision	건의 및 결심
Intelligence prepration for battlefield	전장정보분석
Battlefield area evaluation	전장지역 평가
Terrain analysis	지형분석
Weather analysis	기상분석
Threat evaluation	위협평가
Threat integration	위협종합
Operations plan or order format	작전계획 또는 명령양식
Situation	상황

Enemy forces	적군
Friendly forces	아군
Mission	임무
Execution	실시
Intent	의도
Concept of operation	작전개념
Tasks to maneuver units	기동부대 과업
Tasks to maneuver support unit	전투지원부대 과업
Service support	근무지원
Command and control	지휘 및 통신
Annexes	부록
Distribution	배부선

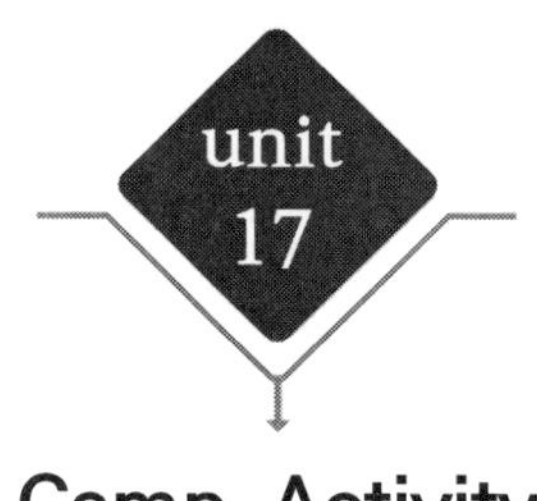

Camp Activity

병영활동

Morning roll call	일조점호
Muster formation	점호대형
Muster call	집합나팔
Bed-check	취침점호
Stand-by inspection	각개점호
Surprise roll call	불시점호
Inspection muster	검열점호
Check roll call	인원점호
Tattoo roll call	일석점호
Night watch; vigilant sentry	불침번
Duty hours	일과시간
Duty officer; officer of the day	당직사관
Barracks leader	내무반장
Military ceremony	군대예식
Flag-raising ceremony	국기게양식
Oath-taking ceremony	선서식
Review, inspection of troops	열병, 사열
Reviewing ground	연병장
Reviewing stand	사열대
Reviewing officer	사열관
Honor guard; ceremonial guard	의장대
Salute battery	예포대
March in review; pass in review	분열
Hand salute	거수경례
Cannon salute	예포경례

Band salute	군악경례
Rifle salute	집총경례
Color salute	군기경례
Color party	기수단
Color bearer	기수
Color guard	기수호위병

Duty & Order

근무 및 명령

Duty	근무
Duty officer	당직사관(사령)
Duty sergeant	당직하사
off duty	비번
on duty	근무중
Officer of the day	일직사관(사령)
Charge of quarters	병사 당번
Special duty	특수근무
Guard duty	위병근무
Commander of the guard	위병사령
Guard	위병
Guard detail	위병 근무병
Officer of the guard	위병장교
Sergeant of th guard	위병조장
Guard house	쥐병소
Guard mount	위병교대
Guard room	위병실
Corporal of the guard	위병조장
Sentry	보초
Sentinel	초병
Sentry post	초소
Food service	급양 근무
Staff duty	일직 근무
Staff duty officer (SDO)	일직장교
Go on sentry	보초근무를 나가다
Keep on the alert	비상경계를 하다

Absent without leave(AWOL)	무단이탈
Post	초소
Relief	교대
Relieve	교대하다, 해체하다
Sentry-box	간이초소
Sentry-go	보초 경계구역
Sentry's orders	보초수칙
Special guard orders	보초 특별수칙
General guard orders	보초 일반수칙
Cannon salute	예포 (경례)
Saluting gun	예포 (포자체)
Escort of honor	의장대
Escort of the color	군기 호위병
Escort of the standard	군기 호위병
Guard of honor	의장병
Minute-gun	조포
Retreat formation	하기식 집합
Review	열병
Parade	분열
Retreat	하기식
Ceremony	의식
Very Important Person (VIP)	귀빈
Take over	인수하다
Give the alarm	경보를 발하다
General order	일반명령
Special order	특별명령
Combat order	전투명령
Operations order	작전명령
Attack order	공격명령
Defense order	방어명령
Movement order	이동명령
March order	행군명령
Administrative order	행정명령
Routine order	일일명령

Military Service

병역

Service status	역종
Eligible conscription status	제1국민역
Disqualified conscription status	제2국민역
Stand-by replacement	제1보충역
Retired replacement	제2보충역
Active service	현역
Service with troops	실병근무
Reservist	보충역, 재향군인
Conscript reserve	보충병
Exemption from military service	병역면제
Conscription system	징병제도
Universal conscription system	국민개병제도
Military service law	병역법
Compulsory service	의무병역
Selective service system	선발병역제도
Voluntary service system	지원병역제도
Service term	복무기간
Recruiting station; Induction station	징병소
Physical examination for draft	징병검사
Conscription year	징집연도
Misfit	부적격자
Draft exemption	징집면제
Temporary draft exemption	징집유예
Determent of exemption	징집연기

Draft determent for family hardships	의가사 징집연기
Draft determent for education	학생 징집연기
Notice of enlistment; draft call	입영통지서, 영장
Voluntary enlistment	자원입대
Hip-pocket order	대기영장
Oath of enlistment	입대선서
Recruit training center	신병교육대
Replacement training center	보충교육대
Slacker	징집기피자
To call up	소집하다
To enlist	입대하다

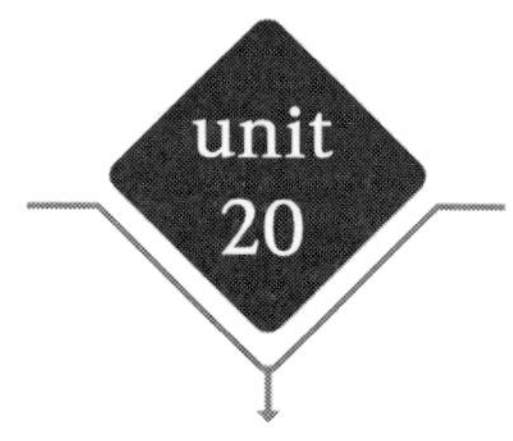

Unit & Quarter Inspection

부대 및 내무검열

Disciplined troops	군기 엄정한 부대
Rules and regulations	제반 규칙
Military justice patrol	군기 순찰
Military justice week	군풍기 강조 기간
Dress inspection	복장 검사
Surprise visit of inspection	불시검열
Unit inspection	부대검열
Post inspection	사후검열
Good appearance	단정한 용모
Military bearing	군인다운 태도
Sense of responsibility	책임감
Joint responsibility	연대 책임
Off-limits to military personnel	군인 출입 금지 구역
Lack of discipline	군기 이완
Service court	군사 재판소
Provost court	계엄 재판소
Court-marshal	군법 회의
Law officer	법무관
Disciplinary committee	징계 위원회
Lockup; detention barrack	영창
Demoralization	군풍기 문란
Breach of discipline	군기 위반
Misconduct	비행
Misdeed	비위 사실
False personation	신분 사칭

Rank misinterpretation	계급 사칭
Self-infliction	자해
Failure in duty; neglect of duty	근무 태만
Misuse of authority	직권 남용
Misuse of leadership	지휘권 남용
Without leave; Without Permission	무단으로
AWOL (absent without leave)	무단 이탈
Desertion	탈영
Desertion from duty	근무 이탈
Insubordination	항명
Mutiny	집단 반항
Usurpation of power by the lower to the high	하극상
Profiting enemy	이적행위
False rumor; canard	유언비어
Submission	항복, 귀순
Official reprimand	징계, 견책
Admonition	훈계
Warming	경고
Corporal punishment	체형
Cookhouse fatigue	취사 근무
Pack drill	완전 군장 얼차려
Disciplinary confinement	근신
Confinement to barracks	외출 금지
Reduction of pay	감봉
Dismissal	파면
Downgrading	강등

Training Program & Test

훈련계획 및 시험

Army Training Program (ATP)	육군훈련계획
Army Training Memorandum (ATM)	육군훈련각서
Master Training Program (MTP)	기본교육계획
Weekly Training Program (WTP)	주간교육계획
Detailed Training Program (DTP)	세부교육계획
Program of Instruction (POI)	교육계획
On the Job Training (OJT)	실무 교육
Army Training Test (ATT)	육군훈련시험
Regiment Combat Test (RCT)	연대전투시험
Reconnaissance, Selection, and Occupation of Position (RSOP)	진지정찰, 선정 및 점령훈련
Maneuver Exercise (MX)	기동훈련
Joint Exercise (JX)	합동훈련
Major Joint Exercise (MJE)	주요합동훈련
Field Training Exercise (ETX)	야외훈련연습
Command Post Exercise (CPX)	지휘소연습
Drill Ground, Training Site, Training Ground	훈련장
Training Schedule	훈련 시간표
Instructor, Drill Master	교관, 조교
Lesson Plan	수업계획
Manual	교범
Field Manual (FM)	야전교범
Technical Manual (TM)	기술교범
Training aids	교육보조재료
Training Equipment	훈련장비

Training Film	교육용영화
Army Training and Evaluation Program (ARTEP)	육군훈련평가계획
Contingency Plans	우발사태 계획
Day and night live fire training	주야간 실사격 훈련
Inclement weather training schedule	악천후 훈련계획
Joint Forces Training	제병합동 훈련
Operation overlay	작전 투명도
Review and analysis	심사분석
ROK/US combined operations	한미 연합작전

PART Ⅱ

Useful Expressions

UNIT 22. Conference I
UNIT 23. Conference II
UNIT 24. Office
UNIT 25. Telephone Conversation I
UNIT 26. Telephone Conversation II
UNIT 27. Restaurant & Party
UNIT 28. Personnel
UNIT 29. Intelligence
UNIT 30. Operations
UNIT 31. Operations Plan
UNIT 32. Logistics
UNIT 33. Engineer
UNIT 34. Communications
UNIT 35. Policy Planning
UNIT 36. Public Affairs

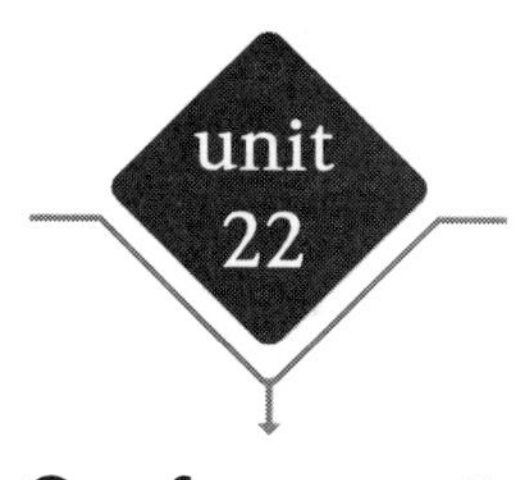

Conference I

회의 I

■ 의견에 찬성(동의)할 때

당신의 계획에 동의합니다.	I agree to your plan.
당신의 계획에 전적으로 동의합니다.	I entirely agree with your plan.
그 계획에 전적으로 찬성합니다.	I'm all for that plan.
그의 의견을 지지합니다.	I support his opinion.
그것을 지지합니다.(재청합니다)	I second that. I support it.
근본적으로 당신의 말에 동의합니다.	Basically, I agree with what you said.
그 점에 대해서는 당신에게 동의합니다.	I agree with you on that point.
그의 의견에 무조건 찬성합니다.	I agree with him unconditionally.
좋은 착상입니다. 찬성합니다.	That's a good point. I agree.
개인적으로 그 제안에 찬성해요.	Personally, I agree to the proposal.
그 계획엔 찬성입니다.	I'm for that plan.

저 역시 그렇게 생각합니다.	I think so, too

■ 의견을 말할 때

제 입장을 말씀드리면 --	As for myself,
제 생각은 --	If you ask me---, I believe---
제는 이렇게 생각합니다.	This is[Here is] what I think.
제 의견은 다음과 같습니다.	In my opinion---
그것에 대해 제 의견이 있어요.	I have my own opinion about that.
제게 다른 의견이 있습니다.	I have another idea[opinion].
좀더 궤적으로 토의할 필요가 있습니다.	We meed to discuss that in a little more detail.
제안이 있습니다.	I have a suggestion.
그것에 대해 검토해 보지요?	Why don't we look into it?
그들의 제의를 받아들입시다.	Let's accept their offer.

■ 표결에 부칠 때

그것에 대해 투표를 합시다.	Let's take a vote on it.
이 일에 대한 최종 결정은 다수결로 정하겠습니다.	The majority will make the final decision.
찬성[반대]하시는 분은 손을 들어 주십시오.	Those in favor[object], please raise your hand.

■ 의견에 반대할 때

저는 그렇게 생각하지 않습니다.	No, I don't think so.
저는 그렇게 여기 않습니다.	No, I don't suppose so.
반대합니다.	I disagree.
당신의 말에 반대합니다.	I disagree with what you are saying.
당신의 의견을 지지할 수 없어요.	I cannot support your opinion.
그 점은 당신과 뜻을 같이 할 수 없습니다.	I can't go along with you on that point.
제 견해는 좀 다릅니다.	I don't see things that way.
그건 그렇게 간단하지 않아요.	It's not that simple.
그건 절대 무리입니다.	That's absolutely impossible!
개인적으로 그 계획은 비현실적이라고 생각합니다.	Personally, I think the plan is unrealistic.

■ 완곡하게 반대할 때

저는 그것에 대해서는 잘 모르겠습니다.	I don't know about that.
글쎄, 어떨지 모르겠습니다.	I'm not sure about that.
이 문제에 대해 다시 한번 생각하고 싶습니다.	I'd like to think twice about this.

■ 회의를 미칠 때

오늘 회의는 이것으로 마칩니다.	Today's meeting is now over.
오늘은 이것으로 끝냅니다.	It's time to end the meeting for today.
다음 회의 때까지 이 건에 대해 검토해 주십시오.	Please think about this before the next meeting.

■ 그 밖의 표현

어떻게 생각하십니까?	What do you think (about(of))---? What is your opinion? How do you like---? Do you have an opinion regarding-?
문제를 좀더(다음에) 의논하는 게 어떻습니까?	Can we discuss this further(another time)?
그건 좀 지나친 것 같습니다.	You are carrying this too far.

Conference Ⅱ

회의 Ⅱ

■ 회의 진행

여기를 주목해 주시겠습니까?	Could I have your attention, please?
회의를 시작하고자 합니다.	I'd like to get the meeting started.
의제로 들어갑시다.	Let's get down to the issue.
본 의제로 들어갑시다.	Let's me get down the main issue.
본론으로 되돌아갑시다.	Let's get back to the main issue.
다음 주제로 넘어갑시다.	Let's go on to the next subject.
질문이 있으시면 손들어 주세요.	Raise your hand if you have any questions, please.
질문을 명확히 해주십시오.	Would you please restate the question more clearly?

■ 질문할 때

질문이 좀 있는데요.	I'd like to ask you something.

질문 있습니다.	Question! I have a question.
질문해도 될까요?	May I ask you a question?
좀도 상세히 설명해 주시죠?	Could you be more specific?
조금만 천천히 말씀해 주시겠습니까?	Would you speak a little bit slower?
그전 슬라이드를 다시 보여 주시겠습니까?	Would you please reshow the previous slide?

■ 의견을 구할 때

당신 생각은 어떻습니까?	What's your opinion?
그의 제안을 어떻게 생각하세요?	What do you think about his proposal? What do you say about his proposal?
그것에 대한 당신의 견해는요?	What are your views on that?
이것에 대해서 당신 생각을 듣고 싶습니다.	I'd like to hear your ideas on this.
잠시 주제에서 벗어난 이야기 좀 해도 되겠습니까?	May I digress (a bit)? May I talk about something off the topic for a moment?
뭔가 말씀해 보십시오.	Please say something.

한동안 생각해 봤지만 좋은 생각이 나지 않는군요. 당신은 어떻습니까?

I've been thinking about it for a while, but I can't come up with any good ideas. How about you?

이 건에 대해 뭔가 제안이 있습니까?

Do you have any suggestions on this?

이 프로젝트에 찬성합니까?

Are you in favor of this project?

이 계획에 반대합니까?

Do you have any objections to this plan?

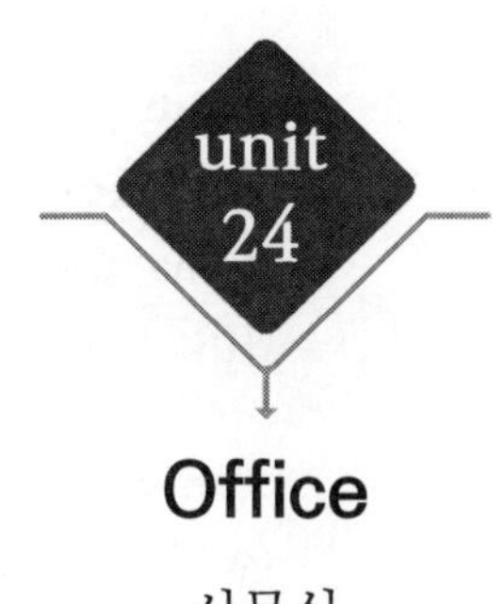

Office

사무실

■ 일을 부탁할 때

심부름을 해주시겠습니까?	Would you please go an errand for me?
도와주시겠습니까?	Could you lend me a hand? Would you do me a favor?
은행에 가줄 수 있겠습니까?	Would you go to the bank for me?
지금 사무실로 올 수 있어요?	Could you come to my office now?
이 보고서를 복사해 주겠어요?	Could you make photocopies of this report?

■ 일의 기한을 물을 때

지금 바쁩니까?	Are you busy [in a hurry]?
급합니까?	Is it urgent?

언제까지 해 드리면 되겠습니까? | How soon do you need to have it done? How soon do you want me to finish it?

마감이 언제입니까? | When is it due? When is the deadline?

내일까지 해도 되겠습니까? | Can it wait until tomorrow?

■ 서류의 소재를 물을 때

정보 자료는 어디에 있습니까? | Where is the intelligence data kept?

월간 보고서가 어디에 있는지 아십니까? | Do you know where the monthly report is?

문서를 찾고 있는데요. | I'm looking for the document.

■ 일의 부탁을 거절할 때

그건 무리네요. | That's impossible.

지금은 너무 바빠서 할 수 없습니다. | I'm too busy, I can't do it at this moment. I'm too busy, I can't do it now.

할 일이 산더미처럼 많습니다. | There's an overwhelming amount of work to do.

죄송하나 지금은 할 수 없습니다. | I'm sorry, but I can't do it right now.

■ 부서장의 승인을 받을 때

이 보고서에 대한 당신의 승인을 받을 수 있습니까?	May I have your approval on this report?
여기 싸인 좀 해주시겠습니까?	Would you please sign here?
내일까지 해주시면 감사하겠습니다.	I'd appreciate having it by tomorrow.

■ 부서장에게 보고할 때

중대한 일이 일어났습니다.	Something urgent has come up, sir.
부대의 사기가 최근에 높아 졌습니다.	Unit morale has gone up recently.

■ 전언(傳言)을 알릴 때

외출하신 동안 몇 가지 전달 사항을 받아두었습니다.	I received some messages while you were out. I took some messages for you while you were gone.
1시간 전에 스미스씨로부터 전화가 있었습니다.	Mr. Smith called you an hour ago. You had a call from Mr. Smith an hour ago.
책상 위에 메모가 있습니다.	There's a memo on your desk.

■ 팩스를 받아 두었을 때

지상작전과로부터 방금 팩스가 왔습니다.	A fax message from Ground Operations Branch has just come in.
방금 당신에게 팩스가 왔습니다.	A fax for you has just come in.

■ OA 기기가 고장 났을 때

복사기가 작동을 안 합니다.	The copy machine doesn't work.
팩스기가 고장 났습니다.	The fax machine is out of order.
프린터가 막 고장 났습니다.	The printer just broke down.

■ 그 밖의 표현

이번에는 아주 열심히 하겠습니다.	I'm going to work wery hard this time. I am going all-out on this.
오늘 오후에 해야 할 일이 많습니다.	There are a million other things I'd rather do this afternoon.
그렇게 하는 것이 당연합니다.	That's the way it's supposed to be.
별 문제 없습니다.	I don't suppose the world would come to an end. I don't have any problems.
금방 돌아오겠습니다.	I'll be back in a flash. I'll be back soon.

일거리가 계속 쌓이는 군요	My workload keeps piling up.
아직 끝나지 않았습니다.	I'm still not done.
모든 일이 잘 될 것입니다.	Everything will work out just fine.
당신이 처한 입장을 충분히 이해합니다.	I really understand where you're at.
저에게 자세히 말해 주십시오.	Fill me in. Please explain to me clearly.

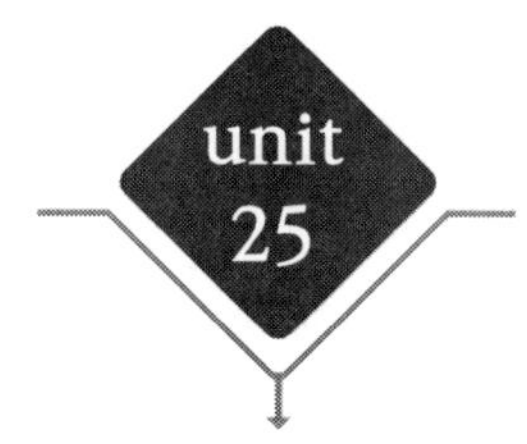

Telephone Conversation Ⅰ

전화통화 Ⅰ

■ 전화 관련 표현

전화를 사용해도 되겠습니까?	May I use your phone? Let me use the phone.
이 근처에 공중전화가 있습니까?	Is there a pay phone around here?
공중전화가 어디 있습니까?	Where can I find a pay phone?
전화를 쓰고 있는데요.	I'm on the line
제게 전화를 걸어 주시겠습니까?	Would you mind calling me back later? Will you please give me a call? Give me a ring(phone, buzz).
끊지 말고 잠깐 기다립시오.	Hold on(hang on) (a minute). Hold(Stay on) the line. Will you hold on a moment, please?
누구를 바꿔 드릴까요?	Who are you calling?

■ 성명을 밝힐 때

저는 홍(길동) 병장입니다.	Hello? This is Sergeant Hong speaking. I'm Sergeant Hong. My name is Hong Kil-dong
인참부 행정과 장 중사입니다.	This is C-1 Admin Sergeant Jang Speaking.
바로 접니다.(전화 바꿨습니다.)	This is he(she) (speaking). It's me. Yes, Kil-dong speaking.

■ 상대를 부를 때

김 소령님 좀 바꿔 주실 수 있습니까?	May I speak to Major Kim? I'd like to speak to Major Kim? Can I speak to Major Kim, please?
전화하신 분은 누구시죠?	May I have your mane, please? Who's calling, please? Who is this, please? To whom am I speaking?
이 소령님 계십니까?	Could you get a hold of Major Lee? Is Major Lee available now? Is Major Lee there(in)?

■ 전화를 받을 수 없을 때

그는 잠시 밖에 나가서 없습니다.	He's not in at the moment. He's out.

그는 지금 전화를 받을 수 없습니다.	He's not available right now.
그는 금방 나갔습니다.	He has just stepped out.
그는 정오까지는 없을 것입니다.	He'll be gone until noon.
그는 다른 전화를 받고 있습니다.	He's talking on another line. He is on the other phone.
나중에 다시 전화 주시겠습니까?	Would you call me again later?
그가 당신에게 전화하도록 할까요?	Shall I have him call you back?
다른 사람을 연결시켜 드릴까요?	Shall I put you through to someone else?

Telephone Conversation Ⅱ

전화통화 Ⅱ

■ 메시지를 요청할 때

제 메시지를 남겨 주시겠습니까?	Can[Could, May] I leave a message(for him)? I would like to leave a message. Will you take a message(for him).
전하실 말씀 있으세요?	May[Can] I take[have] a message? Would you leave(me) a message? Would you care to leave a message? Is there any message?
그거 곧 돌아올 겁니다. 제가 메시지를 전할까요?	He should be back soon, shall I give him a message?
잠시만요, 몇 가지 쓸게요.	Just a moment, please. Let me get something to write on [I need to write it down].
전화 줘서 감사합니다. 그가 메시지를 받도록 확실히 전하겠습니다.	Thank you for call. I'll make sure he gets your message.

■ 전화 응답

네게 전화 왔어.	You are wanted on the phone. Major Kim, telephone. You have telephone call.
전화 받아라.	Answer the phone.
누구십니까?	Who wants me on the phone? Who is on the line? Who is calling? Who is this?
제가 받겠습니다.	I'll get(take) it.

■ 그 밖에 표현

혼선입니다.	The lines are crossed.
시내 전화(시외전화, 장거리 전화,국제전화)를 하고 싶습니다.	I'd like to make a local call(an out-of- town call, a long distance call, an overseas call).
그럼 이만 끊겠습니다.	I've got to hang up now.
그는 지금 출장 중입니다.	He's on TDY(Temporary Duty) now.
죄송하지만 그분은 뉴욕에 출장 가고 안 계십니다.	I'm sorry, but he's in New York for business.
죄송하지만 그가 계신 곳의 전화번호를 알 수 있을까요?	May I have number where I can reach him?

그가 있는 곳에 연락이 되겠습니까?	Can I contact him at his location? Can I contact him at his destination?
그에게 연락할 방법이 있을까요?	Is there any way I can contact him?
어떻게 하면 당신과 연락할 수 있을까요?	How can I get in touch with you? Is there any way to reach you? How can I contact you?
휴대용 전화를 갖고 있습니까?	Do you have a cellular phone?

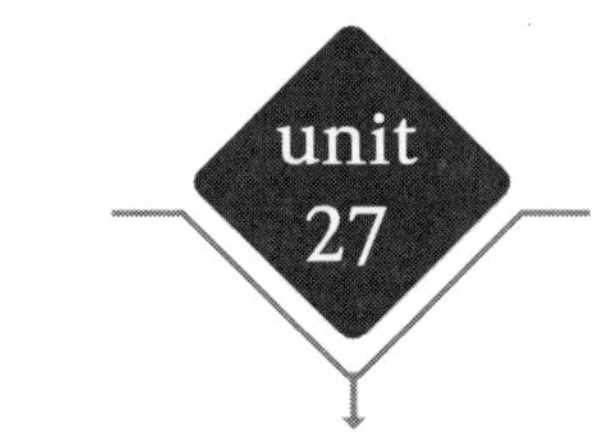

Restaurant & Party

식사와 파티

■ 식당 추천

좋은 식당을 추천해 주시겠습니까?	Could you recommend some good restaurants?
한국 식당이나 중국 식당을 소개해 주시겠습니까?	Could you suggest a Chinese or Korean restaurant?
값이 비싸지 않은 식당을 말씀해 주시겠습니까?	Could you tell me an inexpensive restaurant?
이 부근에 해산물 음식점이 있습니까?	Is there a seafood restaurant around here?

■ 좌석 예약/복장

예약하고 싶습니다.	I'd like to make a reservation, please.
언제 저녁 식사가 준비됩니까?	What time is dinner served?
우리는 6시 반에 자리를 원합니다.	We'd like a table for 6:30. We want a table at 6:30.
2인용 자리를 예약해 두십시오.	Please reserve a table for two?

세 사람이 앉을 자리가 있습니까?	Could we tave a table for three?
창가에 있는 식탁을 원합니다.	We'd like a table near the window.
넥타이를 매야 합니까?	Should I wear a tie?
정장을 해야 합니까?	Do I have to be formally dressed? Is formal attire required?
당신 식당에는 어떤 복장을 해야 합니까?	What is the dress code for your restaurant?

■ 주문 및 음식 취향

메뉴를 볼 수 있습니까?	May I see the menu, please?
주문 받으세요	Will you take my order, please?
지금 주문해도 되겠습니까?	May I order now?
이걸로 주십시오.	May I have this one, please?
같은 것으로 주십시오.	I'll have the same.
How do you like your steak? 로 물으면 4가지 중 하나로 대답합니다.	Rare, please. Medium-rare, please. Medium, please. Well-done, please.
How do you like your coffee? 라고 물으면---	With cream and sugar, please. Just cream, please. Black, please.

How do you like your eggs? 라고 물으면---	Hard-boiled(Soft-boiled), please. Sunny-side up, please. Over easy, please. Scrambled, please. Omelet, please.

■ 식사 중 표현

소금을 건네주시겠어요?	Would our please pass the salt? Would you mind passing the salt? Please pass me the salt.
물 한잔 주시겠어요?	May I have a glass of water?
와인 한 병 주시겠어요?	May I have a bottle of wine, please?
백포도주 두 잔 주세요.	Two glasses of white wine, please.
맥주 주세요.	Beer, please.
제가 건배를 제의해도 될까요? “우리들의 우정을 위하여, 지나간 것들과 앞으로 올 것들에 대하여”	May I propose a toast? "Here's to our friend ship, those in the past, those yet to come."
햄버거 두 개 싸 주십시오.	Two hamburgers to go, please.
이것들을 가져 갈 수 있도록 포장해 주시겠습니까?	Can you bag this for me(us)? Can you wrap this up for me(us)?
포장 주문됩니까?	Can I order a takeout there?

■ 계 산

계산서 주시겠습니까?	Would you bring me the check?
이 계산서 맞죠?	Is this bill correct?
이 신용카드 받습니까?	Do you accept this credit card?
여행자 수표 받습니까?	Do you take traveler's checks?
각자 냅시다.	Let's go Dutch. Let's split the bill. Let's me pay my share. How much is mine? How much is my share?
반씩 냅시다.	Let's go fifty-fifty on the bill. Let's share the bill. Let's go halves.
계산은 제가 하겠어요.	I'll pick up the tab. I'll pay the bill.
점심값은 제가 내겠어요.	The lunch is on me. I'll buy lunch for you.
이것은 무료로 제공하는 것입니다. (술집, 식당 등의 주인이 손님에게 무료 제공시)	This is on the house.
제가 한잔 사겠습니다.	I'll buy a drink.

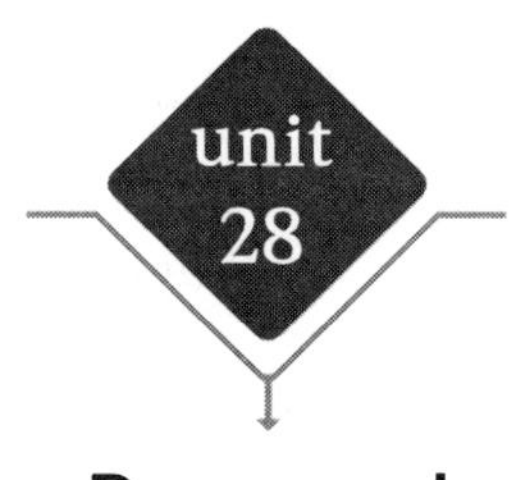

Personnel

인사

보직결정은 누가 합니까? 또 개인 의사는 반영됩니까?	Who decides the duty assignments? Also, do an individual's preferences affect the decision?
병력 보충의 우선 순위는 전술상황 및 병력준비태세 평가에 근거를 둡니다.	Replacement priorities will be based primarily on the tactical situation and personnel readiness evaluations.
이 슬라이드는 지난 12시간 동안의 한미 총 전사상자를 나타내고 있습니다.	This slide indicates ROK and US total casualties over the previous 12hours.
병력현황 보고는 일일 2회 제출하며 06:00 기준보고는 12:00 이전에, 18:00 기준보고는 24:00 이전에 연합사 인사참모부에 도착하여야 합니다.	The Personnel Status Report will be submitted twice daily as of 06:00, to reach CFC C1 NLT 12:00; and 18:00, to reach CFC C1 NLT 24:00.
미군의 사단, 군단 및 전구급 포로 수집시설은 현재 각각 어디에 위치하고 있으며, 수용규모는 얼마나 됩니까?	Where are the collection points, holding areas and temporary EPW camps located? And what are their capacity to hold EPWs?

주한미군 병력현황 정보를 주시겠습니까?	Can you give me information on the current status of USFK personnel strength?
응급환자가 발생하여 긴급 환자 후송용 헬기를 신청하고 싶습니다.	I'd like to request a medical evacuation helicopter to evacuate an emergency patient.
김 일병이 전투 간 어깨 관통상을 입었습니다.	A bullet went through PFC Kim's shoulder during combat.
기수단 통과 시에는 사열대의 주빈을 포함한 모든 사람이 경례를 해야 합니다.	When the color bearer passes the review stand, everyone, including the guest of honor, needs to salute.
행사 시작 5분전까지 참석자는 입장을 완료해 주십시오.	Everyone needs to be in place five minutes prior to the start of the ceremony.
여기 공적조서 양식이 있으니 첨부물을 참조하여 작성해 주십시오.	Here is a form of recommendation for award. Please fill it out according to the attachments.
민간인 참석자의 좌석배치는 어떻게 결정됩니까?	How is the seating arrangement determined for civilian attendants?

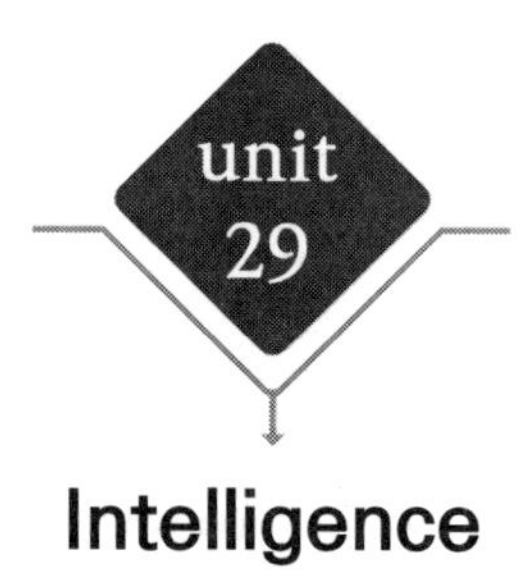

unit 29

Intelligence

정보

어제 저녁 18시경 DMZ에서 북한 특수부대의 활동이 있었습니다.	Last night around 18:00hrs., SOF activities were sighted at the DMZ.
적 특작부대의 주요 타격목표는 무엇이며, 이에 대한 아군측의 대비태세는 무엇입니까?	What are the main targets of the enemy's Special Fores and what kind of preparation do we have against them?
정보감시태세는 몇 단계이며, 단계별 상태와 감시활동 수준은 어떻게 됩니까?	How many levels does WATCHCON have? What activities/events will cause a change in the WATCHCON status?
이번 최초계획회의에서 정보 참모부가 수행해야 하는 역할은 무엇입니까?	What is the role of the G2 section at this initial planning meeting?
이 항공기는 장착하고 있는 감지기는 DMZ로부터 얼마 까지 탐지할 수 있습니까?	How far from the DMZ can the sensor on this airplane detect?
적 후방에서 활동 중인 종심작전 부대에 관하여 설명해 주시겠습니까?	Could you please brief us on the strategic units operation in the enemy's rear area?

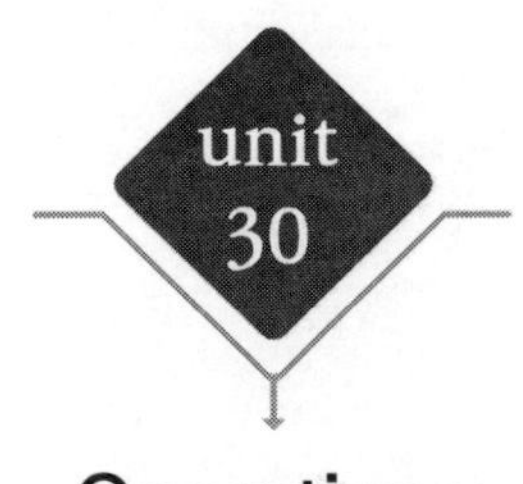

Operations

작전

이 공격의 목적은 포병의 사격진지 변환을 강요하여 사격을 방해하고 지연시키는데 있습니다.

The objective of this attack is to disrupt and delay fires by forcing the systems to alternate firing points.

금일 17시에 보고 예정이니 보고관련된 자료를 주십시오.

I will brief at 17:00 today, so I need you to send me your data.

제가 작성한 영문 스크립트를 문법에 맞게 재작성해 주시겠습니까?

Could you correct the grammar of my English script and rewrite it?

주요 지휘관회의는 12월 13일 서울 지휘소에서 개최될 예정입니다.

The Senior Leaders Meeting will be conducted on 13 Dec in CC SEOUL.

이 사항에 대하여 다시 파악하여 보고 드리겠습니다.

We will get a grip on the situation and report back to you.

부대가 이들 목표를 언제 어디서 달성해야 합니까?

Where and when must our unit accomplish these objectives?

적군은 어떤 형태와 위력의 무기로 아군을 공격할 것이며, 이에 따른 결과는 어떻게 될 것입니까?

What types of weapons will the enemy employ to strike us and what will the results be?

Operations Plan

작전계획

현 상황은 연합위기 상황입니까? 상황이 명백한 북한의 국지도발, 전면전의 징후, 또는 단순한 사건입니까?	Is the situation considered as a combined crisis? Is this incident a deliberate provocation by NK or a prelude to war, or just an accident?
'98년 7월 15일 14:30분경 동해안에서 잠수복을 착용한 북한 무장공비 사체가 발견되었습니다.	The corpse of a north Korean agent in a diving suit was discovered on the east coast around 1430I(India time), 15 July, '98
북한군 항공기가 전술 조치선을 남하 비행함에 따라 공작사에서 13시에서 15시까지 즉각 대응 조치를 하였습니다.	A NK aircraft crossed the TAL, and ACC immediately took counter measures from 1300I to 1500I.
악기상으로 5건의 공중강습 훈련이 최소 되었습니다.	Five AAT missions were cancelled due to bad weather.
보고는 대구 2군사 대회의실에서 아래 계획과 같이 열릴 예정입니다.	These briefings will be held in Taegu, SROKA's main conference room, as scheduled below.
D+2일 목표는 각 여단별 00%를 파괴하는 것입니다.	Our objective is to get 00% destruction of each brigade by D+2.

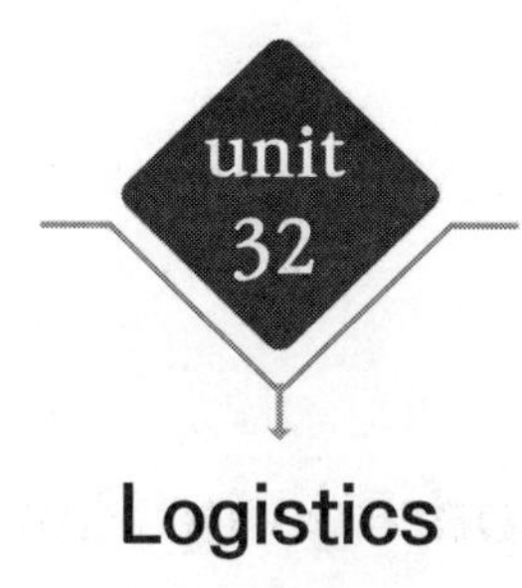

Logistics

군수

최초계획회의의 목적은 연습계획을 수립하며 요구되는 지원과 작전 연습개념을 발전시키는 것입니다.	The purpose of the IPC is to establish exercise planning and to develop the exercise concept of operations and supporting requirements.
사령관 상황보고에 입력할 자료가 없습니까?	Do you have the input for the CINC's SITREP?
미군이 사용하고 있는 보급품에 대한 지급기준은 무엇입니까?	What supply level does the US military use for individual equipment?
미군이 사용 중인 주요 장비와 물자의 특성과 제원을 알고 싶습니다.	I want to know the characteristics and specifications of main USFK equipment and materials.
X사단에 대한 탄약 직접 지원은 어느 탄약 보급소가 담당합니까?	Which ASP will be in charge of the X divisional direct support for munitions?
현대 전체적인 지상군 탄약은 몇 일분입니까?	How many days do munitions last for the entire ground forces?
이번에 실시하는 대화력전 탄약은 어떤 탄약들 입니까?	What kinds of munitions are being used in this counterfire fight?

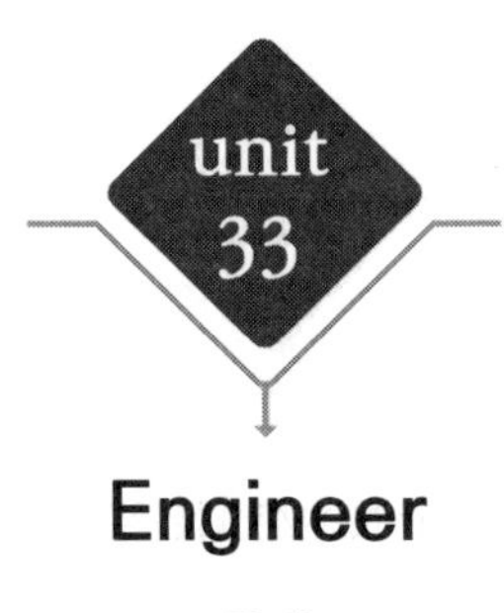

Engineer

공병

공병지원의 우선권은 전방 전투지역 작전지원에 있습니다.	The priority for the allocation of military engineers is to support operations in the FCZ.
전방 전투지역의 공병지원은 기동 및 대기동 그리고 생존 지원에 주안을 둡니다.	The principal objet of the engineering support in the FCZ is to support mobility, countermobility, and survivability
살포식 지뢰는 계획된 장애물의 설치가 시간적으로 어려울 때 운용됩니다.	Scatterable mines are employed when planned obstacles cannot be installed due to shortage of time.
지뢰지대의 종심은 적 부대의 돌파능력을 초과하도록 계획해야 합니다.	The depth of the minefield must be designed to override the enemy forces' ability to breach.
작전과에서는 야전군 및 구성군사공병에 야외훈련 연습 형태 및 일정을 제출할 것을 지시하였습니다.	Operations Branch has ordered the Field Armies and Component Engineers to submit a list of proposed FTX types, dates, and locations.
지형분석은 아군보다 적 지역에 대한 분석이 더 중요합니다.	Terrain Analysis of the enemy area is more important than that of the friendly area.

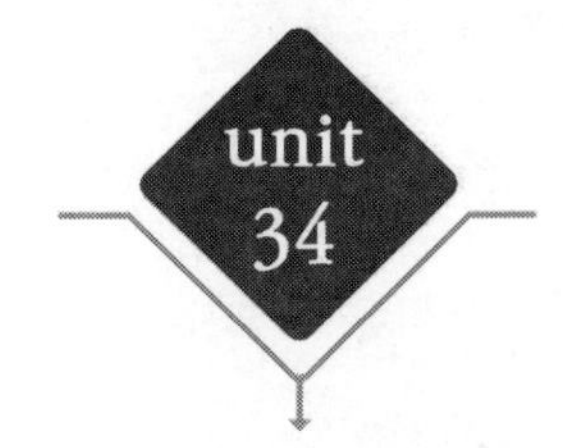

Communications

통신

한국 육군본부 참모들이 미 X통신여단을 방문할 계획입니다. 그들을 위하여 부대소개를 준비해 주시겠습니까?	The staff of ROK Army HQ will visit the US X signal brigade. Would you prepare a mission brief?
XX항공여단의 통신부대 및 장비편성을 알고 싶습니다. 그쪽 담당자와 연락을 하려면 어떻게 하는 지 알려주시기 바랍니다.	I'd like to know the signal unit and equipment organization of XX aviation brigade. How do I contact the action officer?
위기조치실에는 어떠한 통신망이 설치되어 있습니까?	What communications systems are installed in the CAT room?
제 생각에는 패스워드가 잘못된 것 같습니다.	I think the password is invalid.
천둥 번개 시에는 전화선을 설치하지 말아야 합니다.	Do not install telephone wiring during a lightning storm.
현재 우리가 운용하고 있는 국방부 공식 PC용 운영체제는 Windows 95입니다.	Window 95 is currently MND's official operating system for PCs.

Policy Planning

정책기획

우리의 최우선적인 임무는 한반도에서 전쟁을 억제하는 것이며, 억제 실패 시 적을 격퇴하여 한미 양국의 안녕과 번영을 보장하는 것입니다.	Our mission, first and foremost, is to deter war on the Korean Peninsula, and if that deterrence fails, to defeat the enemy and secure the safety and prosperity of out countries.
명일 브리핑은 연합민사사령부 및 연합 훈련에 관한 내용에 중점을 둘 것입니다.	Tomorrow's briefings will focus on information related to the Combined Civil Affairs Task Force and combined training.
연합군사령관은 전쟁억제와 방어 임무를 담당한 전구급 전투사령관입니다.	As Commander in Chief of Combined Forces Command, he serves as the theater level war fighter with missions of deterrence and defense.
유엔군사령관은 공동경비구역을 통제하고, 정전체제 유지를 위해 유엔사 인원을 운용합니다.	CINCUNC employs UNC personnel to control the UNC-joint security area and assist him in maintaining the armistice.

민사작전은 전쟁 이후에만 적용되는 것이 아니라 작전개시전 또는 작전기간 중에도 수행됩니다.

Civil Affairs operations are not just a post-hostilities activity, but operations conducted prior to and during operations as well.

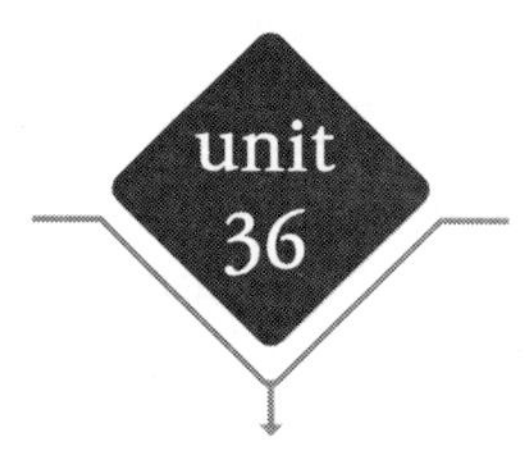

Public Affairs

공보

이번 전시증원연습에 대한 보도지침은 무엇입니까?	What is the Public Affairs Guidance for the RSOI exercise?
언제 보도자료를 발표할 예정입니까?	When are you going to make an announcement of the exercise?
유엔사 장성급회담에 관한 공식적 보도자료를 발표할 예정입니까?	Are you going to make an official statement regarding UNC General Officers Meeting?
오늘의 주요 뉴스는 무엇입니까?	What are today's headlines?
연합상륙훈련에 대해 언론사에서 취재요청이 있었습니까?	Is there any media request to cover the Combined Amphibious Assault Exercise?
취재진에 대한 용산기지 출입조치가 완료되었습니까?	Is the media's arrangement to access Yougsan complete?
좀 더 자세한 정보 입수 시 보도자료를 발표할 예정입니다.	When we have further information, we will make a news release.
요즘 언론의 주요관심사는 무엇입니까?	What's the media's current issue?

주한미군 정보참모부, 법무참모부와 이 서류에 관해서 상의했지만 완전하게 작업할 충분한 시간이 없었다는 것을 이해하여 주시기 바랍니다.

Although we have discussed the document with the USFK J-2 and Judge Advocate Office, please understand that we did not have sufficient time to accomplish full setting.

한미 행정협정 대한민국의 형사 재판권은 전시에 정지되고 미군측이 미군이 저지른 모든 범죄에 관한 단독 재판권을 가지게 됩니다.

The criminal jurisdiction article of the US-ROK SOFA is suspended in the event of war, and US military authorities have exclusive jurisdiction over all offenses committed by US service members.

이 행사는 한국전쟁 50주년을 기념하는 국가행사입니다.

This is a national event commemorating the 50th anniversary of the Korean War.

한국전쟁의 교훈과 역사 및 유산에 대해 더욱 명확한 이해와 인식을 대중에게 제공해야합니다.

We provide the public with a clearer understanding and recognition of the lessons, history, and legacy of the Korean War.

군정위는 쌍방이 각각 임명한 5명으로 구성된 유엔사와 북한군의 공동조직체로 편성되었습니다. 이들 중 최소한 3명은 장군급이어야 합니다.

The MAC was designed as a joint UNC-KPA organization, with five officers from each side. At least three of these officers have to be of flag rank.

1953년 7월 27일에 서명된 한국 정전협정은 오늘날 까지도 지속되고 있는 휴전을 성립시켰습니다.	The Korean Armistice Agreement, signed on July 27, 1953, established a cease-fire which continues to this day.
나는 국제연합군 총사령관으로부터 조선인민군 최고사령관에게 정식 항의를 전달하라는 명령을 받았습니다.	CINCUNC has ordered me to pass a formal protest to the Supreme Commander of the Korean People's Army.
당신측의 모든 비난은 다국적 조사반에 의해서 이미 조사가 완료되었으며 모두가 허위였습니다.	All of your charges have been investigated by multinational team, all findings were untrue.
50명을 넘지 않는 보도인원들이 제 450차 군사정전위원회 회의를 취재하기 위해 1999년 4월 1일에 공동경비 구역으로 들어갈 것입니다.	Not more than 50 press personnel will enter the JSA on 1 April 1999 to cover the 450th MAC meeting.

PART Ⅲ

Examples of Operations Orders

UNIT 37. Company Attack
UNIT 38. Company Defense
UNIT 39. Armored Division Attack
UNIT 40. Infantry Division Defense

Company Attack

중대 공격

The following is sample operation order as it would be issued orally by commanding officer, Company A, 1st Battalion, 66th Infantry. Paragraph titles, numbers, and letters ate shown only for convenience of reference. The company commander would concurrently point out terrain features, including objectives, on a map, sketch, and/or on the ground(fig B-1).

Task Organization: No change except one battalion AT squad is attached under company control, and one squad from 1st Platoon attached to 2d Platoon for assault. (Example of cross attachment within company.)

1. SITUATION

a. Enemy Forces: Estimated enemy rifle platoon in zone. Unit is improving its position. Enemy capable of employing nuclear weapons is zone. Indications are that enemy eill continue to defend.

b. Friendly Forces

(1) TF 1-66 attacks 190500 April to seize high ground vicinity HILLS 528-413.

(2) B/1-66 attacks 190500 April on out left to seize Objectives 8. A 5 min. artillery and mortar preparation on Objectives 1 and 2 will be fired commencing H-05; fires for Objectives 3 and 4, on call.

(3) 1-67 Infantry attacks 190500 April on our right to seize Objective 167.

c. Attachments and Detachments: Task organization

2. MISSION. COMPANY A ATTACKS 190500 APRIL TO SEIZE HILL 401 AND HILL 413; CONTINUES ATTACK NORTH, ON ORDER.

3. EXECUTION

a. Concept of operation: 2d Platoon makes the main attack on right, to seize Objective 1 and, on order, Objective 3. 3d Platoon makes supporting attack on left to seize Objective 2 and, on order, Objective 4. 1st Platoon, in reserve, follows 2d Platoon, by bounds, on order. A 5 min. artillery and mortar preparation on Objective 1 will be fired commencing H-05; fire for Objective 3 on call. Priority of fires to main attack.

b. 1st Platoon:

(1) Be prepared to assist in seizure of OBJ 1 or 3, on order

(2) For consolidation, be prepared to occupy HILL 394, on order.

c. 2d Platoon:

d. 3d Platoon:

e. Mortar Platoon:

(1) GS rear of HILL 381.

(2) Displace on order.

f. Battalion AT Squad:

(1) GS vicinity HILL 391.

(2) Provide AT protection to right flank.

(3) Displace to HILL 398 upon seizure of Objective 1.

(4) Displace to Objective 4 upon seizure.

g. Battalion AD Team:

(1) DS Company A vicinity HILL 381.

(2) Provide AD protection, emphasis in assault forces.

(3) Displace to HILL 420 upon seizure of Objective 1.

(4) Displace to Objective 4. upon seizure.

h. Coordination Instructions: Base platoon for assault is 2d Platoon.

Final coordination line for Objective 3 and 4 is RED Trail.

4. SERVICE SUPPORT

a. Supply and services. Breakfast 0330 on position. One c-ration meal issued per man with breakfast meal. Riflemen carry one extra bandoleer of ammunition. Company trains located rear of HILL 381.

b. Medical Evacuation. Company aid post in woods near RJ 274 displace to rear of HILL 381.

5. COMMAND SIGNAL

a. Signal:

(1) SOI Index 9-12 IN EFFECT.

(2) Emergency signal to left supporting fire is one RED SMOKE STREAMER.

b. Command: I will be with the 2d Platoon. CP moves to rear of Objective 3 upon seizure. Time is now ________. Are there any questions?

Company Defense

중대 방어

Task Organization : No change except battalion AT squad detached to battalion control and one engineer squad is attached and will be under company control.

1. SITUATION

a. Enemy forces. Enemy continues to counterattack 5,000 meters to our east. The enemy which previously occupied HILL 401 and HILL 413 is now believed 1,00 meters to our front. Enemy patrols have been contacted 500 meters to our fiont.

b. Friendly Forces:

(1) TF 1-66 defends along the forward slope of the high ground form HILL 528 on the left to HILL 413 on the right.

(2) B/1-66 defends on our left along the forward slope of HILL 528. C/1-66 in reserve.

(3) 1-67 Infantry defends on our right in the vicinity of HILL 395.

(4) One squad of engineers is attached.

c. Attachments and Detachments: Task organization.

2. MISSION

Company defends by 191400 April on line along the forward slope of HILL 401 and HILL 413.

3. EXECUTION

a. Concept of Operation. We will defend employing two platoons on line: 3d platoon on the left form Red Trail to the Saddle between HILL 401 and HILL 413. 2d platoon from the saddle to the wooded draw between HILL 413 and HILL 395. 1st Platoon will be in reserve in a position to the rear of HILL 413. There is one 105 mm artillery FPF planned to our right front. One heavy mortar FPF is planned to the left front of our area. Priority of artillery and battalion mortars to our company.

b. 1st Platoon: Install an antipersonnel minefield to the front of the 2d Platoon area.

c. 2d Platoon.

d. 3d Platoon.

e. Mortar Platoon: 81mm Mortar Section final protective fires located along the edge of the woods in front of 2d Platoon. 81mm Mortar Section priority of fires to 2d Platoon. AT squads in GS, one squad located in 2d Platoon area.

f. Engineer Squad: Assist 1st Platoon in emplacing minefield.

g. Coordination Instructions. Priority of work: Post security, position weapons and platoons, clear fields of fire, prepare positions, and establish early warming devices. I will assign supplementary positions later.

4. SERVICE SUPPORT

a. Supply and Services. Supper after dark in position. Each platoon have a guide at he company CP by 1710 hours. Company trains are located to the rear of HILL 381. Insure 2,500 rounds of ammunition on position for each machinegun.

b. Medical Evacuation. Company aid post is located in company trains area.

5. COMMAND AND SIGNAL

a. Signal

(1) SOI Index 9-13 in effect.

(2) Emergency signal to fire final protective fires is RED ATAR STAR CLUSTER at night and RED SMOKES STREAMER during daylight.

b. Command. company CP will be located here. Time is now ____ hours. Are there any questions?

Armored Division Attack

기갑사단 공격

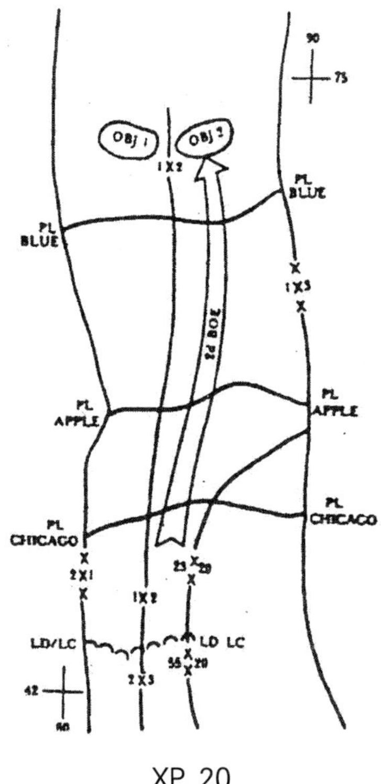

XP 20

(Classification)

Copy No.3 Of ___ copies
23d Armed Div
YUNGCHUN (9107), CHINA
011100 Jun 19____

OPORD 21

Reference: Map, CHINA, 1:50,000, FUKIEN Sheet.

Task organization:

1st Bad	2-Bad
1-91 Mech	1-93 Mech

(Classification)

(Classification)

1-92 Mech	1-94 Mech
1-10 Armor	1-12 Armor
1-11 Armor	1-13 Armor

(OPORD 21-23D Armd Div)

1-50 Arty	1-51 Arty(DS)
A/1-440 Arty	B/1-440 Arty
A/23 Engr	B/23Engr
Sec/E/23d Engr	Sec/E/23d Engr
1/23d MP CO	2/23d MP CO
3d Bde	Div Trp
1-95 Mech	1-22 Cav
1-14 Armor	1-440 Arty (C/V)(-)
1-15 Armor	23d Engr (-)
C/23d Engr	5535th Engr Co (Fltbg)
	(atch eff 011600 Jun)
Div Arty	
	2239th MID (remains atch)
1-52 Arty	23d Sig
1-53 Arty	213th CA Co (remains atch)
1-54 Arty	23d MP Co(-)
2d Bn (155,Sp), 631st Arty	
(atch eff 011400 June)	DISCOM

1. SITUATION

a. Enemy forces. Annex A, Intelligence.

b. Friendly forces

(1) 1st corps atk 020330 June with two divisions in the

(Classification)

(Classification)

assault, 23d Armd Div on the west (left) and 20th Inf Div on the east (Right), secures CHANGCHOW (5080), and prep to cont atk to the morth.

(2) 55th Mech Div follows and supports 23d Armd Div.

(OPORD 21-23d Armd Div)

(3) 201ST Armd Cav Regt protects corps west flank.

(4) Elements 9th TAF support 23d Armd Div.

(5) 2d Bn (8, SP), 671st Arty, reinf 23d Armd Div Arty.

c. Attachments and Detachments. Task organization.

2. MISSION

Division atk 020330 Jun secures crossing over TSIN River, prep to cont atk to the north.

3. EXECUTION

a. Concept of Operation.

(OPORD 21-23D Armd Div)

(1) Maneuver. Division atk thought 55th mech Div with 1st Bad on west, 2d Bad on east, 3d Bde in reserve; secures crossings over STIN River between HUTON (4483) and ANKI (4500) inclusive; prep to cont atk to the north.

(2) Fires. Nuclear fires will ve employed to neutralize Aggressor reserve vic area of pass and against targets of opportunity. A-30 minute nonnuclear artillery preparation will be fired beginning at H-20 minutes. Annex B, Fire Support.

b. 1st Bad

c. 2d Bad

d. 1-22Cav: Follow 2d Bde initially; protect division east flank.

c. Arty:

(1) FA:

(a) 1-50 Arty: DS 1st Bad.

(Classification)

(Classification)

(b) 1-51 Arty: GS, prep DS 3d Bad

(c) 1-52 Arty: GS.

(d) 1-53 Arty: GS.

(f) 1-54 Arty: GS.

(g) 1-631 Arty: GS.

(OPORD 21-23D Armd Div)

(2) ADS: 1-440 Arty (C/V) (-); protect in priority 1-54 Arty, and division HQ

(3) Annex B, Fire Support.

f. 23d Engr Bn (-): Annex C, Engineer.

g. Div Trp: Follow 1st Bad.

h. DISCOM: Remain in present position initially.

i. reserve: 2d Bad: Fol 1-22 Cay during passage, thereafter 2d Bad; prep to assume mission of 1st Bad of 2d Bad; prep to protect division east flank.

j. Coordination Instructions.

(1) Lead brigades coordinate passage of lines zone.

(2) Annex D, Road Movement Table.

4. ADMINSTRATION AND LOGISTICS

Annex E. Administration.

5. COMMAND AND SIGNAL

Annex F. signal. Index 1-3, SOI. Cont listening silence until H-20 minutes.

Acknowledge,

BLANDFORD
MG

(Classification)

(Classification)

Annexes: A-Intelligence (omitted)
B-Fir Support (omitted)
C-Engineer (omitted)
D-Administration (omitted)
F-Signal (omitted)

Distribution: A

(OPORD 21-23D Armd Div)

OFFICIAL.

/s/ McDonald
MCDONALD

63

Infantry Division Defense

보병사단 방어

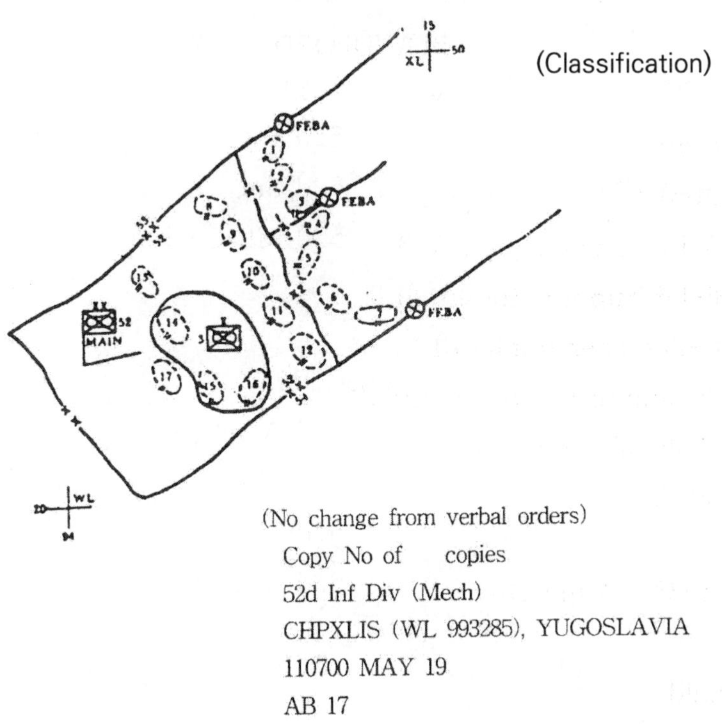

(Classification)

(No change from verbal orders)

Copy No of copies

52d Inf Div (Mech)

CHPXLIS (WL 993285), YUGOSLAVIA

110700 MAY 19

AB 17

OPORD 19

Reference: Map YUGOSLAVIA, 1:50,000. SISAK-IVANIC GRAD Sheet.

Task organization:

1st Bad　　　　　　　　　Div Arty

(Classification)

(Classification)

1-76 Mech 1-42 Arty
1-77 Mech 1-43 Arty
(OPORD 21-23D Armd Div)
1-4 Armor 1-44 Arty
1-40 Arty 2d Bn (8.SP), 610th Arty (remains atch)
A/1-441 Arty 2d Bn (8.SP), 610th Arty (remains atch)

2d Bde
1-78 Mech
1-79 Mech
1-5 Aromr
1-41 Arty (DS)
B/1-441Arty
B/52d Engr(DS)

Div Trp
D/1-23 Cav
1-441 Arty (C/V) (-)
52d Engr
52d Sig
52d MP Co

3d Bde
1-80 Mech
1-81 Mech
1-82 Mech
1-6 Armor
1-23 Cav (-)
C/52d Engr (DS)

DISCOM
52d Admin Co
52d Maint
52d Med
52d S & D

1. SITUATION

a. Enemy Forces. Annex A, Intelligence.

b. Friendly Forces.

(1) 3d (US) Corps defends immediately along SAVA River in sector from ZAGREB (WL 780700) exclusive to BISTRAC (XL 1272270)

(Classification)

(Classification)

inclusive and prepares to continue the attack to the northeast.

(2) 53d Mech Dive is defending on SAVA River in vicinity of DESNA MARTINSAK VES (XL 077496) not later than 111100 May.

(3) 54th Mech Div is defending on SAVA River in sector.

(OPORD 21-23D Armd Div)

(4) Elements 9th (US)TAF support 52d Mech Div.

(5) 72d Arty Gp GSR 52d Mech Div Arty.

c. Attachments and Detachments. Task organization.

2. MISSION

Division defends immediately along SAVA River form DESNA MARTINSKA VES (XL077496) exclusive to XL156310 to hold the enemy northeast of Highway 201.

3. EXECUTION

a. Concept of Operation

(1) Maneuver. Division conducts a mobile defense in sector, employing the 1st Bde and 2d Bde from north to south in the forward defense area. 1st Bde will defend in sector to retain the high ground immediately adjacent to the SAVE River. 2d Bde will delay in sector. 3d Bde, as division res, counterattacks to destroy enemy in penetration; priority of employment to 2d Bde sector.

(2) Fires. Priority of air and artillery fires to 2d Bde inittally; to 3d Bde when committed, Annex B, Fire Support.

b. 1st Bde:

(1) Defend in sector.

(2) Prepare in priority blocking positions 2,1,3.

(Classification)

(Classification)

c. 2d Bde:

(1) Delay in sector.

(2) Prepare in priority blocking positions 5,4,7.

d. D/1-23 Cav:

(1) Maintain contact with 53d Mech Div until unit closes on SAVA River vicinity DESNA MARTINSKA VES.

(2) On order, conduct surveillance of division rear area.

(OPORD 21-23D Armd Div)

e. Arty:

(1) FA:

(a) 1-40 arty: DS 1st Bde.

(b) 1-41 Arty: DS 2d Bde

(c) 1-42 Arty: GSR 1-41 Arty: DS 3d Bde when committed

(d) 1-43 Arty: GS

(e) 1-44 Arty: GS.

(f) 2-601 Arty: reinf 1-41 Arty.

(g) 2-635 Arty: GSR 1-40 Arty.

(2) ADA: 1-441 Arty (C/V) (-): Protect in priority division res, DISCOM, and division CP.

(3) Annex B, Fire Support.

f. 25d Engr:

(1) GS, priority of effort to 3d Bde.

(2) Annex C, Barrier.

(3) Annex D, Engineer.

g. DISCOM: Be operational in new area vic XL 110950 not later then 112400 May.

(Classification)

(Classification)

h. Res: 3d Bde:

(1) Prepare to execute division counterattack plans, priority to 2d Bde sector. Annex E, Counterattack Plans.

(2) Prepare in priority blocking positions 6, 11, 10, 12, 9, 8, 15, 14, 16, 13, 17.

(OPORD 21-23D Armd Div)

(3) Prepare to release up to one- battalion task force on 30-minute to division control for rear area security missions.

I. Coordinating Instructions.

(1) Complete reorganization and preparation for defensive mission by 1111200 May.

(2) Brigades in forward defense area coordinate reconnaissance and security measures forward of the FEBA.

4. ADMINISTRATION AND LOGISTICS

Annex F, Administration.

5. COMMAND AND SIGNAL

Index 1-6, SOI. Annex G. Signal

Acknowledge.

TRACKS

MG

Annexes: A- Intelligence (omitted)
B- Fire Support (omitted)
C- Barrier (omitted)
D- Engineer (omitted)
E- Counterattack Plans (omitted)

(Classification)
(Classification)
F- Adminstration (omitted)
G- Signal (omitted)
H- Air Defense (omitted)
Distribution: A
72d Arty Gp

(OPORD 21-23D Armd Div)
OFFICIAL:
/s/ Carrier
CARRIER
G3

APPENDICES

A. Glossary(English-Korean)
B. Glossary(Korean-English)
C. Special Expressions
D. Abbreviations and Terms

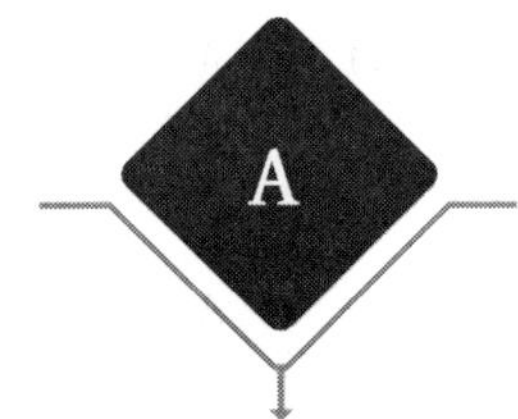

Glossary (English-Korean)

군사용어(영어-한글)

360-degree security	사주경계
AA	집결지
AA (assembly area)	집결지
abatis	목책
About, face!	뒤로 돌앗!
acknowledge	수신여하; 귀소수신 여부
ADA	방공포병
ADA(Air Defense Artillery)	방공포병
adjacent commander	인접부대장
adjacent unit	인접부대
adjoining unit	인접 부대
adjust the rifle	소총(혹은 가늠자)을 조정하다
adjutant general	부관
administrative and/or logistic support	행정 및 군수 지원
advance to be recognized	(확인받기 위하여) 앞으로 가라
after-action report	전투 결과보고; 교전 결과보고
aidman	위생병; 의무병
air assault	공중강습
Air Defense Artillery	방공포병
air interdiction	항공차단
Air Land Battle	공지전투
air liaison officer	공군연락장교
air liaison officer(ALO)	공군연락장교
air observation	공중 관측
airborne	공수
airborne aircraft	비행중인 항공기
airspace	공역(空域)
airstrike	공습, 대지공격
align the sights	조준선 정열을 하다
all or part of	~의 부분 또는 전체
all the rest	기타 모두

all the way	내내;(~에서,~까지) 온통; 쭉
allocate	할당(배분)하다
allocations	배급량
all-round defense	사주방어
ALO	공군연락장교
alternate	예비(대); 교체 부대
alternating roles	번갈아 가면서; 임무 교대하면서
ambush	매복하다(lie in wait)
ammunition bearer	탄약수
annoyance	교대
antiarmor team	대전차조
APC	장갑 인원 수송차
APC (armored personnel carrier)	장갑 인원 수송차
area of responsibility	책임지역
area target	지역표적
armed forces	(육해공군을 망라한) 군대
arming range	장전거리
armor	기갑
armored personnel carrier	장갑 인원 수송차
armored vehicles	장갑차량
army	군
ARTEP	Army Training and Evaluation Program
As far as I know	내가 아는 한
as soon as possible	가능한 빨리
as to	~에 대하여
ASAP	가능한 빨리
ASAP (as soon as possible)	가능한 빨리
aside	방백으로 (혼자 중얼거리는 소리로)
assault	돌격
assault phase	돌격단계
assault position	돌격진지; 돌격지점
assembly area	집결지
asset	자산
assigned or attached crew-served weapon	예속 혹은 배속된 공용화기
assigned sector	할당구역
At ease!	쉬어!
AT-4	다목적 경(輕)대전차회기 (light multi purpose antitank weapon)
attach	배속시키다
attachment and detachment	배속 및 파견

attack heading	항공기의 공격방향
attack position	공격대기지점
attacking echelon	공격 제대
attend	맡다; 책임지다
Attention!	차렷!
audacity	과감성
audible	가청의
augmented	증강된
automatic rifle	자동소총
avenue of approach	접근로
award and decoration	상훈
axis of advance	전진축
ballistic	방탄(의)
barrel	총열
base of fire	사격기지; 기본화력 지원부대; (역습시) 기저부 사격
base platoon	기준소대
basic arm	기본화기
basic training	기초(군사)훈련
battalion	대대
battalion commander	대대장
battalion in line	대대횡대
battery	포병포대
battery comman der	포대장
battle position	전투진지
battlefield survival	전장생존
bayonet	총검; 대검
bayonet drill	총검술 훈련
bayonet training	총검술 훈련
be accustomed to	~에 익숙하다
be alerted for	인지하다
be attached to	~에 배속되어 있다
be beaten	패배하다
be better off	~보다(형편이)낫다
be capable of	~울 할 수 있다
be clear of	~에서 벗어나다
be commensurate in size with	규모상 ~에 상응하다
be depressed	사기가 죽다
be detailed for	~에 임무분담을 하다
be dug in	참호 속에 배치되다
be entrenched	참호진지를 구축하여 방어편성을 하고 있다
be exhausted	지쳐버리다
be fed up	질리다; 싫증나다
be fit for	~에 적합하다
be fond of	~를(을) 좋아하다
be in full swing	한창 진행중이다
be in high spirits	사기가 높다; 기분이 좋다

be in proper condition	양호한 상태에 있다
be off duty	비번이 되다; 근무가 끝나다
be on board a troopship	군용 선박에 승선하다
be on guard duty	위병근무를 서다
be on hand	현재 보유하고 있다
be on one's feet	(병 후에)원기를 회복하다
be referred to as	~라고 불리다
be stuck with	헤어나지 못하다; 계속하다
be that as it may	어쨌든(anyway)
be used to	~에 익숙하다
beat feet	서둘러 가다; 뛰다 (run quickly; hurry)
Behave yourself.	점잖게 굴어라(행동하라).
believe it or not	믿거나 말거나
biological defense	생물학 병어작전
blow the bugle	나팔을 불다
bother	괴롭히다
bottleneck	애로
bound	경계를 정하다
boundary	전투지경선
bracket fire	협차사격
branch	병과
breach	돌파
breach site	돌파지역
breach the obstacle	장애물을 개척하다
break	짧은 휴식
break contact	접촉을 이탈하다
bridging equipment	가교장비
broad jump	넓이뛰기
BT	전투진지
BT(battle position)	전투진지
BTRs	batteries(포대)
buddy	친구
buddy teams'	(기편성이 아닌 현장에서 편성한) 2~3명 단위로 짝지은 조
bugle	나팔
built-up area	건물지역
bullet	탄환
bulletin board	게시판
bull's eye	(표적의)중심; 흑점
bunk	침대
burlap	위장망(굵은 삼베로 만듦)
burst of questions	연속적인 질문
butt	개머리판
by shaking his morale	적의 사기를 뒤흔듦으로써

by surprise	기습으로
cache	(탄약, 식량 등의)은닉장소
call sign	호출부호
call your shot	탄착점을 예언하다
camouflage	위장
camouflage	위장(하다)
canalizing effect	유인효과
canard	유언비어
cannon	대포(평사포, 곡사포, 박격포 등의 총칭)
cannon-cocker	포병인
can't stand	~을 참을 수 없다
capitalize on	~을 이용하다
captain	대위
capture	점령
cardinal direction	기본방위방향
carry on	(하던 일을)계속하라.
cartridge	탄약
cartridge case	탄피
cartridge supply	장전된 탄약
CAS	근접항공지원
case	탄알집
casualties	사상자
Casualty Collection Point	사상자 수집소
casualty evacuation	전상자 후송
CBR (Chemical, Biological, and Radiological)	화생방
CCP (Casualty Collection Point)	사상자 수집소
center squad	중앙분대
ceremonial guard	의장대
CFC (Combined Forces Command)	연합사(연합군 사령부)
chain of command	지휘체계
chain of hills	연달은 고지들
challenge	수하(하다)
challenge and password	수화 및 암구어
charging handle	장전 손잡이
Check Point	확인점
check point (CP)	확인점
checkpoint	확인점
cheer up	원기를 북돋우다; 사기를 돋우다
chemical agent	화학작용제
chemical corps	화학
Chemical, Biological, and Radiological	화생방
chemical, radiological. smoke and flame operations	화학, 방사능, 연막 및 화염작전

chew someone out	꾸짖다; 고함지르다 (yell at someone)
chow call	식사집합 나팔신호 (mess call)
chow time	식사시간 (meal time)
Class Ⅰ	1종 보급품 (음식)
Class Ⅳ	4종 보급품 (공병)
class of supply of interest	관심 보급품의 종
cleaning	손질
clear approach	(방해,지장 등이)전혀 없는 접근
clock reference	시계방향
close	(~에) 접근하다; 근접전투를 실시하다
close(with)	(~에) 접근하다; 근접전투를 실시하다
close air support (CAS)	근접항공지원
close combat	근접전투
close with	~에 접근하다
closed contour line	폐쇄등고선
CO (commanding officer)	지휘관
code word	음어
cohesion	응집력; 결집력; 단결심
cohesive element	통합부대
cold	완전히
collection point	수집소
collocate	인접 배치하다
column	종대
column of companies	대대종대
Column of twos to the left, march!	(이열 종대로) 줄줄이 좌로 갓!
combat arms	전투병과
combat patrol	전투정찰(대)
combat service support	전투근무지원
combat service support arm	전투근무지원병과
combat service support element	전투근무지원부대
combat support	전투지원
combat support arms	전투지원병과
combat support company	전투지원 중대
combat survival	전장생존
combatant	전투원
Combined Forces Command	연합사(연합군 사령부)
Com on!	자!; 어서!
command	부대(a unit)
command	구령; 명령
command a good view of	~를 잘 보다

command and control	지휘 및 통제
command post	지휘소
commanding officer	지휘관
commercial line	민간 전화
commercial phone	민간 전화
company commander	중대장
company executive officer	부중대장
company executive officer (XO)	부중대장
company train	중대 치중대
complete order	완전명령
composition	(부대의) 구성
COMSEC	통시보안
concealment	차장(은폐)
concept of operations	작전개념
concurrently	동시에
considerations	고려사항들
consolidate	진지강화하다; 통합하다
consolidation	모아 합치다
construction material	진지강화
Contact Point	접촉점
contingency	우발상황
contour line	등고선
control measures	통제방책
coordinated	약정된
coordinates	좌표
coordination	협조
corps	군단
corps of engineers	공병
correction	정정
corrective action	교정조치
counter	대응하다
counterattack	역습
course of action development	방책발전
cover and concealment	엄폐와 은폐
covered by	엄호 하에
CP	확인점
CPT	대위
CPT(captain)	대위
cracking of shots	탕탕하는 총소리
cratering streets	도로에 저지공 (대화구)을 만들다
crawling	포복
cross level	균등분배하다
crosswise	가로로
crouching position	지향 사격자세

CS and CSS units	전투지원 및 전투근무지원 부대
CSS	전투근무지원
CSS (combat service support)	전투근무지원
cultural feature	인공물
cultural obstacle	(장애물 중) 인공지물(인공장애물과 구분됨)
cycle of operation	작동순환
DASC	직접항공지원본부
dayroom	휴게(오락)실
deadly effect	치명적 효과
deception	기만
decisive point	결정적 지점
decontamination	제독제
decontamination kit	제독키트
defeat in detail	각개격파
defective parts	결함부품
defensive operations	방어작전
deficiency	부족; 결점
defilade	차폐
defiles	애로
degree of incline	경사도
degree of surprise	기습의 정도
delaying forces	지원(작전)부대
deliberate attack	정밀공격
deliberate crossing	정밀도하
delineate responsibility	책임을 명시하다
deliver fire	사격을 가하다
demolition	폭파
demolitions	폭약
deny	거부(차단)하다
deploy	전개하다; 산개하다
detachment left in contact	잔류접촉분견대
detachment left in contact (DLIC)	잔류접촉분견대
detail	사역 분담을 하다
detainee	역류자
detention barrack	영창
develop the situation	상황을 발전시키다
deviation	이탈
difficult terrain	험란한 지형
digit by digit	숫자별로; 숫자 하나씩
dining hall	식당(군대, 공장 따위 집단의 큰 식당은 mess hall이라고도 함)
direct air support center(DASC)	직접항공지원본부
direction of attack	공경방향(화살표)

discus	원반
disengage from the enemy	적으로부터 이탈하다
disengagement	전투이탈
dismiss	해산하다
dismounted and mounted avenues of approach	보도 접근로 및 차량 접근로
disorganized	와해된
dispensary	의무실
dispersion	산개; 소산
disposition	(부대의) 배치
disrupt	와해하다
disruption	와해
distinguish oneself	두각을 나타내다
division	사단
DLIC	잔류접촉분견대
Double time, march!	뛰어갓!
draft call	입영통지서, 영장, 징집영장 (draft notice)
draft notice	징집영장
Dragon	휴대용 대전차 공격 미사일(M47)
draw	(보급을) 받다; 얻어내다
Dress right, dress!	우로 나란히!
drill ground	연병장
drop and give 10	팔굽혀 펴기를 10회 하다 (do 10 pushups)
duck down	(맞지 않으려고) 홱 몸을 굽히다
duty officer	당직사관
EA (engagement area)	교전지역
earliest time of move	최기 출발시간 (earliest start time, EST)
easily recognized feature	식별이 용이한 지형지물
echelon	제대
effective range	유효사거리
eliminate	제거하다
embankment	제방; 둑
emplaced munitions	비치된 탄약
encode	암호화하다
enemy ambush force	적 매복부대
Enemy Prisoner of War	적 포로
enforcement	집행
engage the enemy	적과 교전하다
engagement area	교전지역
engineer river-crossing company	공병 도하 중대

engineer river-crossing equipment	공병 도하 장비
EPW (Enemy Prisoner of War)	적 포로
escarpment	급경사면
espionage	간첩(행위)
essential task	필수과업
estimate of the situation	상황판단; 상황평가
e-tool	야전삽
evacuation of wounded	부상자 후송
execution	(작전의) 시행
exert	발휘하다; 행사하다
existing obstacle	기존장애물(한국군 교리의 자연장애물에 해당)
exploitation	전과확대
FA(field artillery)	야전포병
FAC	전방항공통제관
facilitate	용이하게 하다
failure in duty	근무 태반
fall in	집합하다
Fall out!	헤쳐!
false rumor	유언비어
false duty	사역; 잡역
FATOC	야전군 전술작전본부
feed a new round	새로운 탄환을 장전하다
feel at ease	안심하다
feel fit	(몸 상태가) 좋다
field army tactical operations center (FATOC)	야전군 전술작전본부
field artillery	야전포병
field latrine	야전변소
field training	야외훈련
fields of fire	사계
fighting morale	전투사기
fighting withdrawal	교전간 철수; 전투간 철수
figures	숫자로 부르겠다
fill	성토; 적토부
final coordination line	최종협조선
finance	경리
finish the enemy	적을 격멸하다
fire and maneuver	사격과 기동
fire integrated support team	통합화력지원반
fire power	화력
fire support	화력지원팀
fire support coordination center (FSCC)	화력지원협조본부
fire support officer	화력지원장교
fire support team	화력지원

fire team	화력팀(분대는 2개 fire team으로 구성됨)
firing line	사선
firing pin	공이
firing practice	사격훈련; 사격연습
firm grasp	확실한 장악능력
first sergeant	중대 행정보급관
first-aid man	위생병; 구급요원
FIST (fire integrated support team)	화력통합지원조
fit into	~에 끼어들다
five paces apart	5보 간격으로
fix	고착(固着)시키다 (pin down)
fix the enemy	적을 고착시키다
fixed target	고정표적
flash deflector	소염기
flash hider	소염기
flash suppressor	소염기
float	(화물 운반용의) 대차(臺車)
flu	유행성 감기 (influenza) 독감(cold)
FO(forward observer)	전방관측자
FO RATELO (forward observer RATELO)	전방관측소 무전병
follow-in forces	후속부대
foothold	발판; 기지
footlocker	사물함
for an indefinite time	무기한으로
forbid	금지하다
fording	도섭
formation	(정지간) 제식훈련; 집합(대형)
fortification	축성
forward air controller(FAC)	전방항공통제관
forward observer	전방관측자
forward observer RATELO	전방관측소 무전병
forward position	전방 진지
forward slope	전사면
foxhole	개인호
FPF	Final Protective Fire (최후방어사격)
fragmentary order	단편명령
fragmentation grenade	세열수류탄

FRAGO (fragmentary order)	단편명령
form the scene	그 자리로부터
front sight	가늠쇠
FSCC	화력지원협조본부
FSO(fire support officer)	화력지원장교
full colonel	대령
gain a decision	승리하다
general and special guard orders	일반 및 특별 보초 수칙
geographic feature	지형물
get hit badly	심한 타격을 받다
get into trouble	곤경에 빠지다; 골치를 앓게 되다
get shot	사격을 받다; 사격을 당하다
get up	기상하다
Give me a hand ...	날 좀 도와다오; 좀 도와줄래
give the alarm	경보를 발하다
go ahead	계속 말하라
go ashore	상륙하다
go light infantry	경보병으로 가다
go on sick call	환자집합하다; 수진하러 가다
go over	복습하다
go to bed	취침하다
government issue	관급물품
grazing fire	최저표척사격
grenade	유탄(hand grenade와 rifle grenade가 있음)
grenadier	유탄발사기 사수
gird square	격자방안
grid system	방안 좌표법
grin	히죽이 웃다
ground observation	지상관측
ground of vital importance	중요지형
ground tactical plan	지상전술 계획
guard duty	위병근무(보초, 순찰 등의 근무)
guardhouse	위병소
gym(gymnasium)	체육관
Halt!	제자리 섯!
hammer	공이치기
hand grenade	수류탄
handle	취급하다
hand-to-hand fighting	백병전
hasty attack	급속공격
hasty defensive position	급조 방어 진지
hasty protective minefield	급조 방호지뢰지대

heavy stands of trees	빽빽하게 서있는 나무들
heavy weapons company	중화기중대
heliport	헬리콥터 발착장; 헬기장
help oneself to	~을(마음대로 집어) 먹다
high jump	높이뛰기
hinge on	~에 달려있다
hit the mark	표적에 명중시키다
honor down	고착시키다
honed	연마된
honor guard	의장대
hoof it	걷다(go on foot)
hostile territory	적지
house the firing mechanism	격발장치를 수용하다
hump	짊어지다(등 위나 어깨에 메다)
I read back	복창한다
I say again	재송한다
I spell	음성문자로 송신한다
I verify	확인한다
IAW(in accordance with)	~에 따라(서)
identification	신분 확인
identify	식별하다
ignite	점화시키다
illumination	조명
immediate area	인접지역; 지근지역
in a single file	일렬종대로
in accordance with	~에 따라(서)
in an advisory capacity	자문관 자격으로
in charge of	~을 담당한
in light of	~을 고려하여; ~에 비추어
In place, double time, march!	제자리로 뛰어갓!
in style	멋있게
induced limited visibility	인위적으로 조성된 제한된 시계
induction center	모병(징병)소
induction station	징병소
Infantry	보병
infantry, air assault, airborne rifle platoon	보병, 공중강습, 공수소총소대
infiltration	침투
initial attack formation	최초공격대형
initiative	주도권
inspect	검열하다
inspecting officer	검열관
inspection in quarters	내무검열
inspection in ranks	부대검열
installation	시설물

instruction	지시; 교육
integrate	통합하다
intent	의도
internee	구류자
issuance	하달
issue	불출하다, 지급하다
jamming	전파 방해
javelin	투창
jeopardize	위태롭게 하다
jeopardizing target	위협적인 표적
Keep ~ in good shape	~를 양호한 상태로 유지하다
keep from getting lost	방향을 잃지 않게
keep on the alert	빈틈없는 경계를 하다
keep step	발을 맞추다
key terrain	중요 지형(지물)
key terrain feature	중요지형지물
kill zone	살상지역
kitchen police	취사근무; 취사병
knock into shape	(고된 훈련으로) 군인답게 만들다
knock it off	하고 있는 행동을 멈추다; 동작 그만 (stop talking; shut up)
knock off	제거하다
KP (kitchen police)	취사근무; 취사병
lame duck	절음발이 오리
land navigation	독도법
landmark	지표(물)
lateral maneuver	측방기동
laterally	측면으로
LBE	(탄약, 무기 등)휴대용 장비
LC	접촉선
LD	공격 개시선
LD time	공격 개시시간
lead element	선도부대
lead squad	선두분대
leading rifle element	선도소총부대
leave something you of	~에서 ~을 제외하다
left flak squad	좌측분대
Left-as you were!	좌로-다시!
len itself to	~에 적합하다
level crossing	(철도) 건널목
lieutenant (LT)	중/소위
lieutenant colonel (LTC)	중령
lifesaver	인명구조대원
lifting of fire	사격연신
lifting or shifting	(사격) 연신 또는 전환
light condition	경하상태(輕荷狀態)

light infantry rifle platoon	경보병 소총소대
likely avenues of approach of	예상접근로
line	횡대
line of companies	대대횡대
line of contact (LC)	접촉선
line of departure (LD)	공격 개시선
line of duty	근무계통
link-up	연결
linkup point	연결점
listening silence	무선침묵
litter evacuation	(부상자용) 들 것을 이용한 후송
live ammunition	실탄
load management	전투 휴대량 관리
load-bearing equipment(LBE)	(탄약, 무기 등)휴대용 장비
loading and locking	장전 및 잠금
locate	찾아내다
locate a weakness	취약점을 찾아내다
locker	정돈함
lockup	영창
lubrication	주유
lull	일시적인 전투 중지; 소강상태
lump	혹
machine gun nest	기관총좌
magazine	탄알집; 탄창
Maggie's drawers	감적호에서 표적에 안 맞았을 때 흔드는 적기(赤旗)
main attack	주공
main effort	주공; 주력
maintenance	정비
make a rush	약진하다
make one's bed	침구 정돈을 하다
malfunction	기능장애
maneuver	기동
maneuver unit	기동부대
map reading	독도법
map reconnaissance	지도정찰(기상 또는 항공정찰의 예비단계)
march in review	분열
marked	현저한; 두드러진
marker's gallery	감적호
mask	제한하다; 기리다
mass	집중하다
master sergeant	상사(SFC 중사와 SGM 원사의 중간 계급)
materiel	물자
math	수학 (mathematics)

MAW (medium antitank weapon)	중(中)대전차회기
mechanic	기술자, 기계수리공
mechanized infantry	기계화보병
medic	위생병
medical company	의무중대
medical corpsman	위생병
medical officer	군의관
medical orderly	당직 위생병
medium antitank weapon	중(中)대전차화기
mental agility	정신적 기만성
mess sergeant	취사담당관
military camp	(반 영구적인) 부대 주둔
military crest	군사적 정상
military occupational specialty	군사주특기
military symbol	군대부호
minor wound	경상
mission specific	임무에 특정한
mission-essential task	팔수임무과업
mission-orinted protective posture	임무형 보호태세 (MOPP)
mobile field mess	이동 야전 취사반
mobile land warfare	기동지상전
mobility	기동성
modified table of organization and equipment (MTOE)	수정된 편성 및 장비표
momentum	공격기세
mortar platoon	박격포소대
mortar section	박격포반
MOS (military occupational specialty)	군사주특기
motorized rifle platoon	차량화 소총소대
mount guard	경계를 서다
mounted forces	탑승병력
move by bounds	(구간) 이동하다
movement	(이동간) 제식훈련
movement control pont	이동통제점
movement to contact	접적이동
moving target	이동표적
MRP (motorized rifle platoon)	차량화 소총소대
MTOE	수정된 편성 및 장비표
MTP (Mission Training Plan)	임무훈련계획
multiple lanes	복수침투로

muzzle	총구
natural lines of drift	자연적으로 형성된 이동이 용이한 통로
natural obstacle	(장애물 중)자연지물
navigation	위치확인
NCO (noncommissioned officer)	부사관
neglect of duty	근무 태만
neutralize	무력화하다
next higher command	차상급부대
night vision devices	야간관측장비; 야시장비
night watch	불침번
NLT (no later than)	~까지는
no sweat	괜찮아; 걱정 말아
no wonder	이상할 것이 없다
noise and light discipline	방음 및 소등 군기
noncom (noncommissioned officer)	부사관
nondenomination al service	초 교파적 예배
notice of enlistment	입영통지서, 영장
nuclear and nonnuclear fire	핵 및 비핵 화력
nucleus	핵심; 중심
nullify	무력화하다
oath of enlistment	입대선서
OBC	초등군사반
objective rally point	목표상 재집결지
observation post	관측소
observe	준수하다
OCOKA	지형평가 5개 요소
odd and even numbers	홀수와 짝수
offensive operations	공격작전
officer of the day	당직사관
Officers' Basic Course(OBC)	초등군사반
off-loading	하역작업
old man	직속상관 등을 칭하는 속어
on guard	징계근무중인
one at a time	한번에 하나씩
OP (observation post)	관측소
open area	개활지
open machinegun fire	기관총 사격을 개시하다
open ranks	간격을 넓히다
open space	개활지
operations order	작전명령

operational assistance	작전지원
OPORD (operation order)	작전명령
oral operations, order	구두 작전 명령
ordnance	병기
organization for combat	전투편성
organization record	부대기록
orient	적응시키다
orientation by map and compass	지도와 나침반에 의한 정치(定置)
ORP (objective rally point)	목표상 재집결지
out	교신 끝
outfit	군인의 개인 피복 및 장비 일체
outpost	전초
over	이상
overlay	투명도
overwatch	엄호하다
overwatching platoon	감시소대
pale	창백한
parade ground	연병장; 열병 또는 분열식 장소
pass	외출증; 외박증
pass in review	분열
passage	인계/인수
passage point	통과지점
passive measures	수동적 방책
password	암구어
patrol	정찰(대)
patrol route	정찰로
pay off	성과가 나타나다
peak	정상; 정점 (summit)
penetration	돌파
personal clothing ans equipment	개인 피복 및 장비
personal hygiene	개인위생
personnel management	병력관리
pertinent	관계있는; 적절한
pertinent information	관련첩보
phase line (PL)	통제선
phase the attack	공격을 단계화하다
phonetic alphabet	음성문자
phony minefield	모의지뢰지대
physical examination	신체검사
physical exercise	체력단련; 육체적인 훈련
physical toughness	육제적 강인성
physical training	체력단련; 육체적인 훈련

pick up the radiation	반사능에 오염되다
pile, arms!	걸어 총!
pin ~ down	~를(을) 고착시키다
pistol belt	딴띠
pit	감적호
pitch and strike	(천막따위를) 가설하고 철거하다
pivot	구심점
PL	통제선(Phase Line)
platoon	소대
platoon headquarters	소대본부
platoon sector sketch	소대작전구역 약도
platoon sergeant (PSG)	소대선임하사; 부소대장
playing field	운동장
PMCS requirements	예방정비 점검 및 수리
point man	첨병
point target	점표적
point well taken	잘 지적했다(You pointed it out well); 잘 받아들여졌다; 그 말도 맞다
pole vault	장대 높이뛰기
police	청소하다
position	배치하다
post	(영구적인) 군 주둔지
potable	식(수)용
preassault bombardment	공격준비폭격
preclude	제외하다; 배제하다
preparation phase	준비단계
preplanned missions	기계획 임무
prevention and investigation of crime	범죄 예방 및 수사
primary fighting positions	주전투진지
primer	뇌관
principles of war	전쟁원칙
priority of fires	화력의 우선권
priority of targets	표적우선순위
priority/precedence	우선권/ 우선권내의 우선순위
private	이병
probable line of deployment	예상전개선
procedure words (prowords)	통화 약어
procure	조달하다
projectile	탄환
propellant	추진장약
proposed target	요청표적
protocol and liaison	의전 및 연락

provision	보급품
prowords	통화 약어
PSG	소대선임하사; 부소대장
PT (physical training	체력단련; 체육훈련
pull ~ in	~을(를)잡아당기다
put out	(불, 담배 등을)끄다
qualify	자격사격을 실시하다
quarter	25센트 짜리 미국 동전
quartering party	설영대
quartermaster	병참
radio	무선
radio check	감도점검
radio operator	무전병
radio station	무선통신소
radio telephone operator	무선전화병; 통신병; 무전병
radius fallout	낙진 반경
raid	습격
rally point	집결지
rallying point	재집결 지점
range card	사경도
range-finding works	거리 특정작업
ranger rifle platoon	특공소총소대
RATELO (radio telephone operator)	무선전화병; 통신병; 무전병
rations	전투식량
rattle	딸랑거리는 소리
read back	복창하라
ready to copy	받아 쓸 준비됐나
ready, front!	바로!
rear area operations	후방지역작전
rear sight	가늠자
receiver	몸통
recognition signal	인식표시
reconnaissance company	수색 중대
reconnaissance patrol	수색정찰(대)
reconnoiter	수색하다; 정찰하다
recruit	신병; 초년병
recruiting station	징병소
rectangle	직사각형
reference point	참조점
refine	세밀하게 하다
regiment	연대
registered letter	등기우편
rehearsal	예행연습
reinforcing obstacle	강화 장애물(한국군 교리의 인공장애물에 해당)

release point (RP)	분진점
release the firing mechanism	격발장치를 작동시키다
relief feature	기복지형
relier in place	전투간 현지교체
relieve	교대하다
render	~하게 만들다
reorganize	재편성하다
repair	수리
repel	격퇴하다
replace	교체하다
replacement	후임자; 보충역
report to	~에 출두하다
reposition squads and weapons	분대 및 화기를 제배치하다
required of leaders	지휘자에게 요구되는
resemble	~와 유사하다
reserve	예비; 예비대
rest plan	휴식계획
restrictions	제한사항
retain	유지하다; 확보하다
retirement	철퇴
retreat formation	하기식 집합
retrograde operations	후퇴작전
reveille	기상 점호 집합
reverse slope	반사면(反斜面)
ridge	능선
ridgeline north to	~북쪽 능선
rifle	강선
rifle range	소총사격장
rifle training	소총훈련
rifling	강선
Right face!	우향 우!
right flank squad	우측분대
rigid	엄밀한; 엄격한
road crater	도로 대화구
road junction	도로교차점
roger	수신 양호
rolling	선박의 옆놀이
room for dispersion	소산공간
rounds of ammunition	여러 발의 탄약
route forward	전진로
routing	전달; 발송
RP	분진점
ruck march	배낭행군하다 (ruck = rucksack)
rucksack	배낭
rules of engagement	교전규칙
sabotage	태업
saddle	안부
safety regulations	안전수칙

salute to the colors	국기에 대한 경례
satchel charge	휴대용 폭약 및 기폭 장치
say again	재송하라
scabbard	대검집
scheme of maneuver	기동계획
scout	척후; 척후병; 척후기 (정찰용 비행기)
screen	차장하다(아군의 작전을 적군이 알지 못하도록 함)
section	반(班)
sectors of responsibility	책임구역
secure	확보하다
security	경계; (부대)안전; 보안
security force	경계부대
security perimeter	외주경계
segregated	격리된
seize or secure	~을 탈취 및 확보하다
seizure	탈취
self-propelled combat vehicle	자주전투차량
semiautomatic	반자동의
sensor	감지기
sentinel(광의)	보초
sentry	보초
separate command	독립 부대
sergeant	병장(육군); 부사관의 통칭
sergeant first class	중사(육군)
Sergeant	하사, 중사, 상사 등 부사관을 통틀어 호칭함
service company	근무중대
set out	출발하다
SFC(sergeant first class)	중사 (육군)
SGT(sergeant)	병장(육군); 부사관의 통칭
shifting of fire	사격전환
shipshape	장연한; 정돈된
shock offect	충격 효과
short of	못 미쳐서
shortages and shortcomings	부족 및 결함
shot	총성
shot group	탄착군
shoulder weapon	견착식 화기
show up	나타나다 (appear)
showdown inspection	공개검열; 진열검열
sighting and aiming	조준선 정렬 및 정조준
sights	가늠장치(가늠자와 가늠쇠)

signal corps	통신
signal interference	신호에 의한 통신 간섭
signal operation instructions	통신운용지시
silence	무선 침묵
silence lifted	무선침묵대기해제
single lane-one squad	단일침투로-분대 동시침투
single lane-staggered squad	단일침투로-분대별 순차적 침투
sketch	요도
skirmish line	산개대형
sling	멜빵
sloping plain	경사진 평원
sluggishly	완만하게
smoke	연막
snap	(나뭇가지 등) 딱 부러지는 소리
sniper	저격수
So much for	~에 대해서는 이만큼 해 두자
SOI (signal operation instructions)	통신운용지시
solid ground	딱딱한 땅
something	꽤 가치있는 일
SOP	예규, 내규
sortie	(전투기의) 출격
souvenir	기념품; 선물
speak slower	천천히 송신하라
special detail	특수임무; 특별작업
special-purpose attack	특수목적공격
sped to the rear	(급히) 후송하다
splash	물이 튀기는 소리
spoil	전리품, 약탈품
sporadic	산발적인
sprinting	단거리 (100m~400m) 달리기
sprocket	전차기동륜
squad drill	분대훈련
squad leader	분대장
Squad, attention!	분대, 차렷!
squeeze	당기다
SSG (staff sergeant)	하사(육군, 해병대); 병장(공군)
Stack, arms!	걸어 총!
staff sergeant	하사(육군, 해병대); 병장(공군)
stand to attention	차렷 자세를 하다
stand for	나타내사; 상징하다
stand-down for maintenance	정비를 위한(군사행위의) 일시정지
stand-to	대기; 경계태세
star cluster	오성신호탄
stated mission	명시된 임무

static targer	고정표적
stay-behind operations	잔류작전
stock	개머리
stop short	갑자기 멈추다
stoppage	(기계적) 고장
straddling	협차사격
strength	병력
strike flight	공습비행대
stuff up	가득 채우다
stumble into	모르고 들어가다
subdivision	하위집단
subparagraph	소항목
succession of command	지휘권 승계
supplementary position	보조진지
supply personnel / mess personnel	보급병 / 취사병
supply room	보급 창고(실)
supply sergeant	군수담당관
supporting attack	조공
supporting fire	지원사격
suppressive fire	제압사격
sure object to	반드시 ~를 반대한다
survival technique	생존기법
survive immediate combat operations	당면한 전투작전에서 생존하다
suspender	멜빵
sustainment load	유지휴대(장비 및 물자)
sustainment operations	유지작전
swish	바스락 바스락 스치는 소리
switchboard	교환대
synchronize	통합하다
T-62	T-62형 탱크
TACC	전술항공통제본부
TACSOP (tactical standing operation procedures)	전술작전예규; 전술적 행동 절차
tactical air control center(TACC)	전술항공통제본부
tactical air control parties(TACP)	전술항공통제반
tactical air support element(TASE)	전술항공지원대
tactical control measures	전술통제방책
tactical operations center(TOC)	전술작전본부
tactical road march	전술도보행군

tactical standing operating procedures	전술작전예규; 전술적 행동 절차
tailor	맞게 선택하다
take advantage of	~을 최대로 이용하다
take aim	정조준하다
take on oath	선서하다
take off	(옷 따위를)벗다
take one's temperature	~의 체온을 재다
take over	~을(를) 이수하다; 일을 떠맡다
Take, arms!	풀어 총!
tangle foot	지선
tanker	기갑인
taps	취침나팔: 점호
TASE	전술항공지원대
task organization	편조, 부대편성
tear up	찢어버리다
template	형판; 상황판
tenet	준칙
tentative plan	잠정계획
terrain feature	지형지물
terrain model	사판; 지형모형
theaterwide	전 전역(全戰域)에 걸친
tie in ~ to(with)	~을 ~와 묶다
tied in with an FPF	최후방어사격과 결합되면
tight situation	곤경; 위험한 처지
time	시간을 지정하다; 시기에 맞추다
time of attack	공격시시 기간
time of flight	비과시간
time on target (TOT)	표적공격시간, 폭탄이 표적에 동시에 도작하여 폭발하는 시간
time proven	오랜 시간에 걸쳐 입증 된
T-MRE-T	조식은 T-ration, 중식은 MRE, 석식은 T-ration으로 한다
to and from	오고 가는
breach	제거하다; 개척하다
to good advantage	상당히 유리하게
relieve	교대하다
seize	탈취하다
spot check	불시점검하다
take turns	교대하다
TOC	전술작전본부
toilet kit	세면도구함
toilet article	세면도구(비누, 샴푸 등)
topographic crest	지형적 정상
topographic symbol	지형부호

TOT	표적공격시간, 포탄이 표적에 동시에 도착하여 폭발하는 시간
track	궤도
trafficability	이동가능성 (capability of terrain to bear traffic)
trail	뒤를 따르다
trail squad	후미분대
train	병참부대
trainee	훈련병
training ammunition	교탄
training projectile	교탄
transient target	이동표적
transmit	송신하다
trench	참호
trench line	참호선
trigger	방아쇠
trigger guard	방아쇠 울
trigger housing group	방아쇠 뭉치
triple standard concertina fence	삼열 표준 윤형철조망
tripods for crew-served weapons	공용화기 삼각대
troop safety and operational exposure guidance	부대 안전 및 작전 노출 방지지침
troop-leading procedures	부대지휘절차
turn	차례
two command levels higher	2차 상급 지휘단계
under the command of	~의 지휘하에
unit	단위부대
unity of effort	전투력의 통합, 노력의 통일
unlock	(자물쇠를) 풀다
Unpile, arms!	풀어 총!
until relieved	임무 교대될 때까지
up front	선두(에)
varying ranges	상이한 사거리
vee	알파벳 글자 'V'
vegetation	(집합적) 식물 (plants); 목초
verbatim	(한 마디 한 마디) 말 그대로의; 축어적(逐語的)인
verify	검증하다
vic	vicinity(~부근)
vigilant sentry	불침번
violence	맹렬함; 폭력
visibility	가시성
voice circuit	음성회로

voice transmission	음성전송(통신)
vulnerable flank	취약한 측방
warning order	준비명령
water velocity	유속
weapons squad	화기분대
weapons training	화기훈련
wear	마모
wedge	삼각대 (대형)
well forward	상당히 앞쪽에; 상당히 전진 하여
well-defined terrain feature	명확한 지형지물
what on earth	도대체 무슨 이야기를 ...
what's up?	뭐가 나올지(what will be served)
white glove inspection	흰 장갑을 낀 손으로 청소 상태를 확인하는 검사
Who's there?	누구냐?
Why not?	왜, 못할 리가 있겠어.
wilco	정확히 수신하였으며 지시대로 시행하겠다
wipe out	소탕하다
wire (communication)	유선
wire (obstacle)	철조망
with stealth	은밀하게
withdrawal	철수
withdrawal not under pressure	자발적 철수
withdrawal under pressure	강요에 의한 철수
within sight of	~가 보이는 곳에서
without leave	무단으로
without permission	무단으로
work up	연구하다; 정리하다; 계산하다
XO	부중대장
zero in	영점사격하다; 영점을 획득하다
zone of action	작전지역

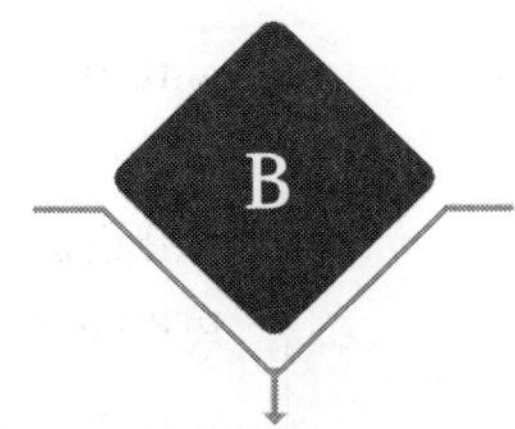

Glossary (Korean-English)

군사용어(영어-한글)

1종 보급품(음식)	Class Ⅰ
2~3명 단위로 짝지은 조(기편성이 아닌 현장에서 편성한)	buddy team
25센트 짜리 미국 동전	quarter
2차 상급 지휘단계	tow command levels higher
4열 종대	column of fours
4열 횡대	line of fours
4종 보급품(공병)	Class Ⅳ
5보 간격으로	five paces apart
가교 장비	bridging equipment
가늠쇠	front sight
가늠자	rear sight
가늠장치(가늠자와 가늠쇠)	sights
가능한 빨리	ASAP(as soon as possible)
가득 채우다	stuff up
가로로	crosswise
가리다	mask
가설반	wire-laying team
가설본부	construction center
가설하고 철거하다(천막 따위를)	pitch and strike
가스 작용식	gas-operated
가시성	visibility
가시신호	visual signals
가청의	addible
각개격파	defeat in detail
각개약진	individual rush
각개점호	stand-by inspection
각개훈련	individual training
간격	interval
간격을 넓히다	open ranks
간이초소, 보초소	sentry box
간접(행위)	espionage
건호병	hospital orderly

간호부장	matron
간호장교	army nurse
갈퀴	extractor
감도 여하	how do you hear me
감도점검	radio check
감독 및 보안하다	supervise and refine
감봉	reduction of pay
감사	plunging fire
감시소대	overwatching platoon
감시초소	look out position
감적수(監的手)	target detail
감적호	pit; maker's gallery
감적호에서 표적에 안 맞았을 때 흔드는 적기(赤旗)	Maggie's drawers
감제고지	commanding heights
감지기	sensor
감찰	IG (Inspector General)
감청반	monitoring team
갑자가 멈추다	stop short
강등	downgrading
강선	rifle; rifling
강습 단정	assault boat
강습 상륙단	assault landing team
강습주정	assault craft
강요에 의한 철수	withdrawal under pressure
강화 장애물(한국군 교리의 인공장애물에 해당)	reinforcing obstacle
개머리	stock
개머리판	butt
개머리판 머리치기	butt stroke to the head
개인 피복 및 장비	personal clothing and equipment
개인위생	personal hygiene
개인호	foxhole
개척하다	to breach
개활지	open space; open area
거리	distance
거리 측정작업	range-finding works
거리목측	judging distance by eye
거부(차단)하다	deny
거수경례	hand salute
걱정 말아	no sweat
건강 정밀 검사	health screening
건강 증명서	health certificate
건강 진단	physical checkup

건강 진단서	bill of health
건널목,(철도)	level crossing
건물지역	built-up area
건설공병	Construction Engineer
건의 및 결심	recommendation and decision
걷다(go on foot)	hoof it
걸어 총!	Stack, arms!; Pile, arms!
검사 총!	Inspection, arms!
검역소	quarantine station
검열관	inspecting officer
검열점호	inspection muster
검열하다	inspect
검증하다	verify
게릴라 작전지역	guerrilla warfare operationa larea (GWOA)
게릴라전	guerrilla warfare
게시판	bulletin board
격리된	segregated
격발	percussion
격발장치를 수용하다	house the firing mechanism
격발장치를 작동시키다	release the firing mechanism
격자방안	grid square
격퇴하다	repel
견착식 화기	shoulder weapon
결점	deficiency
결정적 지점	decisive point
결집력	cohesion
결함부품	defective parts
경계	security
경계근무중인	on guard
경계를 서다	mount guard
경계를 정하다	bound
경계부대	security force
경계분견대	security detachment
경계태세	stand-to
경고	warning
경리	Finance Corps
경보를 발하다	give the alarm
경보병 소총소대	light infantry rifle platoon
경보병으로 가다	go light infantry
경사도	degree of incline
경사진 평원	sloping plain
경상	minor wound
경하상태(輕荷狀態)	light condition
계습 사칭	rank misinterpretation
계산하다	work up
계속 말하라	go ahead
계속하다	be stuck with

계속하라; (하던 일을)	carry on
계엄 재판소	provost court
계획을 완성하다	complete the plan
고고도 방공	high altitude air defense
고려사항들	considerations
고려하여, ~을	in light of
고문	torture
고산지대	alpine terrain
고장,(기계적)	stoppage
고정표적	static target; fixed target
고착(固着)시키다 (pin down)	fix
고착시키다	hold down
고착시키다, ~를(을)	pin ~ down
고함지르다 (yell at someone)	chew someone out
곤경	tight situation
곤경에 빠지다	get into trouble
골치를 앓게 되다	get into trouble
공개검열	showdown inspection
공격 개시선	Line of Departure (LD)
공격 개시시간	time of attack
공격 대기지점	attack position
공격 방향	direction of attack
공격개시 시간	time of attack
공격개시선	line of departure
공격개시시간	LD time
공격기세	momentum
공격대기지점	attack position
공격방향(화살표)	direction of attack
공격을 단계화하다	phase the attack
공격작전	offensive operations
공격제대	attacking echelon
공격준비폭격	preassault bombardment
공군연락장교	air liaison officer (ALO)
공대공 방어	air-to-air defense
공병	Corps of Engineers
공병 도하 장비	engineer river-crossing equipment
공병 도하 중대	engineer river-corssing equipment
공수	airborne
공수부대	parachute troops
공습, 대지공격	airstrike
공습경보	air alert
공습비행대	strike flight

공역(空域)	airspace
공용화기 삼각대	tripods for crew-served weapons
공이	firing pin
공이치기	hammer
공정/공중기동작전	airborne and airmobile operations
공제선	akyline
공중 관측	air observation
공중 기동전술	airmobile tactics
공중 기동훈련	airmobile training
공중 전방항공 통제관	air forward air controller(A-FAC)
공중 전투초계	air combat patrol (ACP)
공중 지휘 통신소	airborne command and communication center
공중강습	air assault
공중회랑	air borne corridor
공지 작전망	air ground operations net
공지 협동	air ground cooperation
공지작전 체제	air-ground operations system (AGOS)
공지전투	Air Land Battle
공포탄	blank ammunition
과감성	audacity
관계있는	pertinent
관급물품	government issue
관련첩보	pertinent information
관심 보급품의 종	class of supply of interest
관측 및 사계	observation and field of fire
관측소	OP(observation post)
관측소 (observation post)	OP
관통된 탄흔	bullet hole
관통상	piercing bullet wound
광명 제원	light data
괜찮아	no sweat
괴롭히다	bother
교대하다	to relieve; to take turns
교대하다	relieve
교두보	bridgehead
교란	annoyance
교신 끝	out
교육	instruction
교전 결과보고	after-action report

교전 간 철수	fighting withdrawal
교전규칙	rules of engagement
교전지역	EA (engagement area)
교정조치	corrective action
교차사격	interlocking fire
교체 부대	alternate
교체하다	replace
교탄	training projectile; training ammunition
교환대	switchboard
구간전진엄호	bounding overwatch
구경	caliber
구급용구	first-aid outfit
구급키드	first-aid kit
구급요원	first-aid man
구두 작전 명령	oral operations order
구령	command
구류자	internee
구성,(부대의)	composition
구심점	pivot
구호소	medical relief station; aid station
국군 의무 사령부	Armed Forces Medical Command
국군통합병원	armed forces hospital
국기게양식	flag-raising ceremony
국기에 대한 경례	salute to the colors
국민개병제도	universal conscription system
군 주둔지,(영구적인)	post
군	army
군기 순찰	military justice
군기 엄정한 부대	patrol
군기 위반	breach of discipline
군기 이완	lack of discipline
군기경례	color salute
군단	corps
군대, (육해공군을 망라한)	armed forces
군대부호	military symbol
군대예식	military ceremony
군법회의	court-marshal
군사 재판소	service court
군사적 정상	military crest

군사정전위원회	Military Armistice Commission (MAC)
군사주특기	MOS (military occupational specialty)
군수담당관	supply sergeant
군수지원	Logistic Support
군악경례	band salute
군용 선박에 승선하다	be on board a troopship
군용 전화	military phone
군의관	medical officer; army doctor
군인 출입 금지 구역	off-limits to military personnel
군인다운 태도	military beating
군인답게 만들다 (고된 훈련으로)	knock into shape
군인의 개인 피복 및 장비 일체	outfit
군종	Chaplains
군풍기 강조 기간	military justice week
군풍기 문란	demoralization
권봉(拳棒)	pudgy stick
궤도	track
귀소수신여부	acknowledge
규모상 ~에 상응하다	be commensurate in size with
균등분배하다	cross level
그 말도 맞아.	Point well taken
그 자리로부터	form the scene
그럼, 깃발은 뭔가	Then, what about flags
근무 복귀	return to duty
근무 태만	failure in duty; neglect of duty
근무가 끝나다	be off duty
근무계통	line of duty
근무이탈	desertion from duty
근무중대	service company
근무지원	service support
근신	disciplinary confinement
근접 전투	close combat
근접수색	close-in search
근접전투	close combat
근접전투를 실시하다	close (with)
근접항공지원	CAS (Close Air Support)
금지하다	forbid
급경사면	escarpment
급속 도하	hasty river-crossing
급속공격	hasty attack
급조 방어 진지	hasty defensive position

급조 방호지뢰 지대	hasty protective minefield
기갑	Armor
기갑보병	Armored Infantry
기갑인	tanker
기갑포병	Armored Artillery
기계화보병	mechanized infantry
기계획 임무	preplanned missions
기관총 사격을 개시하다	open machinegun fire
기관총좌	machine gun nest
기념품	souvenir
기능선택 스위치	selector level
기능장애	malfunction
기동	maneuver
기동 공간	maneuver space
기동계획	scheme of maneuver
기동부대	maneuver unit
기동부대 과업	tasks to maneuver units
기동성	mobility
기동지상전	mobile land warfare
기만	deception
기복 및 배수	relief and drainage
기복지형	relief feature
기본방위방향	cardinal direction
기본화기	basic arm
기본화력 지원부대	base of fire
기분이 좋다	be in high spirits
기상 점호 집합	reveille
기상분석	weather analysis
기상장교	SWO (staff weather officer)
기상지원부대	WSU (weather support unit)
기상첩보	weather information
기상하다	get up
기수	color bearer
기수단	color party
기수호위병	color guard
기술자, 기계수리공	mechanic
기습으로	by surprise
기습의 정도	degree of surprise
기저부 사격,(역습시)	base of fire
기존장애물(한국군 교리의 자연장애물에 해당)	existing obstacle
기준소대	base platoon
기지	foothold
기초(군사)훈련	basic training
기타 모두	all the rest

기후첩보	climactic information
꼬질대	cleaning rod
꽂아, 칼!	Fix bayonets!
꽤 가치있는 일	something
꾸짖다	chew someone out
끄다,(불, 담배 등을)	put out
끼어들다, ~에	fir into
나타나다 (appear)	show up
나타내다	stand for
나팔	bugle
나팔을 불다	blow the bugle
낙진 반경	radius fallout
난청 지역	cone of silence
날 좀 도와다오	Give me a hand...
낫다,~ 보다(형편이)	be better off
내가 아는 한	As far as I know
내내	all the way
내무검열	inspection in quarters
내무반장	barracks leader
내선작전	operations on interior lines
넓이뛰기	broad jump
노리쇠	bolt
노리쇠 전진!	Close chamber!
높이뛰기	high jump
뇌관	primer
누구냐	who's there
능대대형	diamond formation
능선	ridge
능선,~ 북쪽	ridgeline north to
ㄷ 자 대형	three-sided formation
다목적 경(輕)대전차 화기(light multi purpose antitank weapon)	AT-4
단거리(100~400m) 달리기	sprinting
단결심	cohesion
단계	phase
단발 사격	single shot
단위부대	unit
단일침투로-분대 동시 침투	single lane-one squad
단일침투로-분대별 순차적 침투	single lane-staggered squad
단정한 용모	good appearance
단초	single
단편명령	fragmentary order
달려있다, ~에	hinge on
담당한, ~을	in charge of

당기다	squeeze
당면한 전투작전에서 생존하다	survive immediate combat operation
당직 위생병	medical orderly
당직사관	duty officer; officer of the day
대검	bayonet
대검꽂이	bayonet stud
대검집	scabbard
대공 포판	marking panel
대공방어	antiaircraft defense
대공표지	air marking
대기	stand-to
대기영장	hip-pocket order
대대	battalion
대대 무선망	battalion net
대대종대	battalion in column; column of companies
대대횡대	battalion in line; line of companies
대령	full colonel
대상륙 지뢰지대	anti-amphibious minefield
대위	CPT (captain)
대응하다	counter
대전차조	antiarmor team
대차(臺車), (화물 운반용의)	float
대포(평사포, 곡사포, 박격포 등의 총칭)	cannon
대하여, ~에	as to
대해서는 이만큼 해두자, ~에	So much for
도대체 무슨 이야기를...	what on earth
도로 대화구	road crater
도로교차점	road junction
도로에 저지공(대화구)을 만들다	cratering streets
도보 접근로 및 차량 접근로	dismounted and mounted avenues of approach
도보정찰	foot reconnaissance
도섭	fording
도섭능력	fordability
도하 작전	river-crossing operations
도하지점	crossing site
독감	flu
독도법	land navigation; map reading
독립 부대	separate command
돌격	assault
돌격 진지	assault position
돌격단계	assault phase
돌격지점	assault position

돌연 표적	pop-up target
돌출부	salient
돌파	penetration; breach
돌파지역	breach site
동시에	concurrently
동작 그만 (stop talking)	knock it off
동초	flying sentry
두각을 나타내다	distinguish oneself
두격 (경기관총)	head space
두드러진	marked
둑	embankment
뒤로 2보 갓!	Two step backward, march!
뒤로 돌앗!	About, face!
뒤로돌아!	Whirl!
뒤로돌아동작	whirl movement
뒤를 따르다	trail
들 것을 이용한 후송(부상자용)	litter evacuation
들것	stretcher
등고선	contour line
등기우편	registered letter
따라, ~에	IAW(in accordance with)
따라서, ~에	IAW(in accordance with)
딱 부러지는 소리 (나뭇가지 등)	snap
딱딱한 땅	solid ground
딸랑거리는 소리	rattle
뛰다(run quickly)	beat feet
뛰어갓!	Double time, march!
레이저 유도 폭탄	laser guided bomb(LGB)
마모	wear
만들다, ~하게	render
말 그대로의, (한 마디 한 마디)	verbatim
망쳐놓다	spoil
맞게 선책하다	tailor
맡다	attend
매복	ambush
매복하다(lie in wait)	ambush
맹렬함	violence
멋있게	in style
멜빵	suspender; sling
멜빵 조여!	Adjust, slings!
면역 증명서 (예방 접종 증명서)	immunization register
명령	command
명령을 하달하다	issue the order
명시된 임무	stated mission
명확한 지형지물	well-defined terrain feature

모르고 들어가다	stumble into
모병(징병)소	induction center
모아 합치다	consolidate
모의지뢰지대	phony minefield
모의탄	dummy ammunition
목책	abatis
목초	vegetation
목표	objective
목표 재집결 지점	objective rally point
목표상 재집결지	ORP (objective rally point)
몸통	receiver
못 미쳐서	short of
무기한으로	for an indefinite time
무단이탈	AWOL (absent without leave)
무력화하다	neutralize
무력화하다	nullify; neutralize
무선	radio
무선 전파 방해	radio jamming
무선 침묵	(radio) silence
무선통신소	radio station
무선도청	radio intercept
무선연락	radio contact
무선전화병	RSTELO(radio telephone operator)
무선침묵	listening silence
무선침묵대기해제	silence lifted
무전망	radio net
무전병	RSTELO(radio telephone operator); radio man; radio operator
묶다, ~와 ~을	tie in ~ to (with)
문서 보관소	records center
문서 참조번호	message reference number
물이 튀기는 소리	splash
물자	materiel
민간 박명종	EECT (end of evening civil twilight)
민간 박명초	BMCT (beginning of morning civil twilight)
민간 전화	commercial line; commercial phone
믿거나 말거나	believe it or not
밀림	dense vegetation
바로!	Ready, front!
바스락 바스락 스치는 소리	swish
박격포반	mortar section
박격포소대	mortar platoon
반(班)	section

반드시 ~를 반대한다	sure object to
반사면(反斜面)	reverse slope
반자동의	semiautomatic
반장전된	half loaded
받다,(보급을)	draw
받들어 총!	Present, arms!
받아 쓸 준비됐나?	Ready to copy?
발송	routing
발을 맞추다	keep step
발판	foothold
발행부대	issuing headquarters
발행장소	issuing place, country
발휘하다	exert
방공작전	air defense operations
방공포병	ADA(Air Defense Artillery)
방백으로 (혼자 중얼거리는 소리	aside
방사능에 오염되다	pick up the radiation
방아쇠	trigger
방아쇠 뭉치	trigger housing group
방아쇠 울	trigger guard
방안 좌표법	grid system
방어작전	defensive operations
방어준비태세	Defense Readiness Condition (DEFCON)
방음 및 소등 군기	noise and light discipline
방책발전	courses of action development
방책분석	analysis of courses of action
방책비교	comparison of courses of action
방탄(의)	ballistic
방탄조끼	bullet-proof jacket
방향을 잃지 않다	keep from getting lost
배급량	allocations
배낭	rucksack
배낭행군하다	ruck march; ruck march
배부선	distribution
배속 및 파견	attachment and detachment
배속되어 있다, ~에	be attached to
배속시키다	attach
배제하다	preclude
배치	disposition

배치, (부대의)	disposition
배치하다	position
백병전	hand-to hand fighting
번갈아 가면서	alternating roles
번호!	Count off!
번호구령에 없이	without the numbers
번호구령에 맞추어	by the numbers
범죄 예방 및 수사	prevention and investigation crime
법무	JAG (Fudge Advocate General's Corps)
법무관	law offecer
벗다,(옷 따위를)	take off
벗어나다, ~에서	be clear of
변조 방식	modulation method
병과	branch
병기	Ordnance Corps
병력	strength
병력관리	personnel management
병역면제	exemption from military service
병역볍	military service law
병장	sergeant
병장(공군)	SSgt (staff sergeant)
병장(육군)	SGT (sergeant)
병참	Quartermaster Corps
병참부대	train
보급 창고 (실)	supply room
보급병 / 취사병	supply personnel / mess personnel
보급품	provision
보병	infantry
보병	Infantry
보병, 공중강습, 공수소총소대	infantry, air assault, airborne rifle platoon
보안	security
보안감청	security monitoring
보이는 곳에서, ~가	within sight of
보인다	I have ~ in sight.
보조진지	supplementary position
보초	sentinel(광의); sentry
보초경계구역	sentry to
보초수칙	sentry's orders
보초순찰경호	sentry beat
보초특별수칙	special guard orders

보충교육대	replacement training center
보충병	conscript reserve
보충병	replacement
보충역, 재향군인	reservist
보통탄	ball ammunition
복무기간	service term
복수침투로	multiple lanes
복습하다	go over
복장 검사	dress inspection
복좌용수철	recoil spring
복창하라	read back
복창한다	I read back
복초	double sentry
본부통신망	headquarters net
부관	Adjutant General
부근	vic(vicinity)
부대	command; unit
부대 안전 및 작전 노출 방지지침	troop safety and operational exposure guidance
부대 지휘절차	troop-leading procedures
부대검열	inspection in ranks; unit inspection
부대기록	organization record
부대이동을 개시하다	initiate movement
부대주둔, (반영구적인)	military camp
부대지휘절차	troop-leading procedures
부록	annex
부분 또는 전체, ~의	all or part of
부사관	NCO (noncommissioned officer); noncom
부사관의 통칭	SGT (sergeant)
부상병	invalid soldier
부상자 후송	evacuation of wounded
부소대장	platoon sergeant(PSG)
부적격자	misfit
부족	deficiency
부족 및 결함	shortages and shortcomings
부중대장	company executive officer(XO)
분대 및 화기를 재배치하다	reposition squads and weapons
분대, 차렷!	Squad, attention!
분대장	squad leader
분대훈련	squad drill

분열	march in review; pass in review
분진점	release point (RP)
분해결합	disassembly and assembly
불리다, ~라고	be spot check
불발	surprise
불시 점검하다	to spot check
불시점호	roll call
불출	issue
불출하다	issue
불침번	night watch; vigilant sentry
비과시간	time of flight
비밀등급	classification level
비번이 되다	be off duty
비위 사실	misdeed
비정규전	unconventional warfare
비추어, ~에	in light of
비치된 탄약	emplaced munitions
비행	misconduct
비행 협조 본부	flight coordination center (FCC)
비행중인 항공기	airborne
빈틈없는 경계를 하다	keep on the alert
빨치산 조직	partisan organization
빼어, 칼!	Unfix bayonets!
빽빽하게 서있는 나무들	heavy stands of trees
사각지대	dead zone
사격 개시!	Commence fire!
사격 그만!	Cease fire!
사격 금지선	on fire line (NFL)
사격 지원 통제선	fire support control line(FSCL)
사격 통제탑	control tower
사격 협조선	fire coordination line (FCL)
사격과 기동	fire and maneuver
사격구령	fire command
사격군기	fire discipline
사격기지	base of fire
사격명령	fire order
사격연습	firing practice
사격연신	lifting of fire
사격을 가하다	deliver fire
사격을 당하다	get shot
사격을 받다	get shot
사격전환	shifting of fire
사격지휘망	fire direction net
사격훈련	firing practice
사경도	range card
사계	fields of fire

사기가 높다	be in high spirits
사기가 죽다	be depressed
사기를 북돋우다	cheer up
사단	division
사단 지휘망	division command net
사로(射路)	firing lane
사물함	footlocker
사본번호	copy ... of ... copies
사사(斜射)	oblique fire
사상자 수용시설	casualty staging facility
사상자 수집소	CCP (Casualty Collection Point)
사상자(수)	casualties
사선	firing line
사선(射線)	firing line
사수 표적 앞으로!	Firers move down range!
사수, 엎드려 쏴!	Firers, assume the prone
사역	fatigue duty
사역 분담을 하다	detail
사열관	reviewing officer
사열대	reviewing stand
사주경계	360-degree security
사주방어	all-round defense
사판	terrain model
사후검열	post inspection
산개	dispersion
산개 공간	room for dispersion
산개대형	skirmish line
산개하다	deploy
산발적인	sporadic
산악 침투복	rough-terrain suit
산악병	mountain sickness
산악부대	mountain troops
산악작전	mountain operations
살상지역	kill zone
삼각대(대형)	wedge
삼각대(양각대)	tripod (bipod)
삼군	three fighting services
삼열 표준 윤형철조망	triple standard concertina fence
삽탄장전식 (탄창식)	magazine loading
상(하)단막기	high (low) block
상당히 앞쪽에	well forward
상당히 전진하여	to good advantage
상륙군	landing forces
상륙작전	landing operations
상륙하다	go ashore

상사	master sergeant (SFC 중사와 SGM 원사의 중간계급)
상이한 사거리	varying range
상징하다	stand for
상황	situation
상황 및 방책	situation and courses of action
상황분석	analysis of situation
상황을 발전시키다	develop the situation
상황판	template
상황판단	estimate of the situation
상황평가	estimate of the situation
상훈	award and decoration
새로운 탄환을 장전하다	feed a new round
생물학 방어작전	biological defense
생소한 지형	unfamiliar ground
생존기법	survival technique
서둘러 가다	beat feet
선도부대	lead element
선도소총부대	leading rifle element
선두(에)	up front
선두분대	lead squad
선물	souvenir
선박의 옆놀이	rolling
선발병역제도	selective service system
선서식	oath-taking ceremony
선서하다	take an oath
선회지점	orbit point
설대대형	wedge formation
설영대	quartering party
성과가 나타나다	pay off
성토; 석토부	fill
세면도구 (비누, 샴푸 등)	toilet articles
세면도구함	toilet kit
세밀하게 하다	refine
세부임무분석	detailed mission analysis
세열수류탄	fragmentation grenade
세워 총!	Order, arms!
소강상태	lull
소규모 저항지역	pocket
소대	platoon
소대 작전구역 약도	platoon sector sketch
소대본부	platoon headquarters
소대선임하사 (부소대장)	platoon sergeant (PSG)
소사	searching fire

소산	dispersion
소산공간	room for dispersion
소염기	flash deflector; flash
소음기	suppressor; flash hider silencer
소이탄	incendiary ammunition
소집하다	to call up
소총 사격장	firing range
소총(혹은 가늠자)을 조정하다	adjust to rifle
소총사격장	rifle range
소총훈련	rifle raining
소탕하다	wipe out
소항목	subparagraph
손질	cleaning
손질 및 수입	care and cleaning
송신하다	transmit
송탄불능	failure to feed
송탄불량	misfed
수동적 방책	passive measures
수류탄	hand grenade
수리	repair
수색 정찰대	reconnaissance patrol
수색중대	reconnaissance company
수색임무	reconnaissance duty
수색정찰(대)	reconnaissance patrol
수색차장	reconnaissance screen
수색하다	reconnoiter
수송	Transportation Corps
수신 양호	roger
수신여하	acknowledge
수정된 편성 및 장비표	modified table of organization and equipment(MTOE)
수진하러 가다	go on sick call
수집소	collection point
수하	challenge
수하하다	challenge
수학(mathematics)	math
수화 및 암구어	challenge and password
숫자 하나씩	digit by digit
숫자로 부르겠다	figures
숫자별로	digit by digit
쉬어!	At ease!
쉬어자세	relax position
습격	raid
승리하다	gain a decision
시간을 지정하다	time

시계	visibility
시계방향	clock reference
시기에 맞추다	time
시설물	installation
시행, (작전의)	execution
식(수)용	potable
식당	dining hall(군대, 공장 따위 집단의 큰 식당은 mess hall이라고도 함)
식림	cultivated vegetation
식물	vegetation(집합적); plants
식별이 용이한 지형 지물	easily recognized feature
식별하다	identify
식사시간	chow time; meal time
식사집합 나팔신호	chow call; mess call
신병, 초년병	recruit
신병교육대	recruit training center
신분사칭	false personation
신분확인	identification
신체검사	physical examination
신호에 의한 통신 간섭	signal interference
실병근무	service with troops
실시	execution
실탄	live ammunition
싫증나다	be fed up
심한 타격을 받다	get hit dably
아군	friendly forces
안 보인다	No joy.
안부	saddle
안심하다	feel at ease
안전,(부대)	security
안전수칙	safety regulation
알파벳 글자 'V'	vee
암구어	password
암호 운용지시	crypto operation instructions
암호 취급소	crypto center
암호 통신문	crypto correspondence
암호화하다	encode
앞에 총!	Port, arms!
앞으로 가다, (확인 받기 위하여)	abvance to be recognized
애로	defiles; bottleneck
야간관측방비	night vision device
야시장비	night vision device
야외 사격장	field range

야외 훈련	field training
야외훈련	field training
야음차 (밤시간 범위)	nocturnal range
야전 병원	field hospital
야전 위생대	ambulance corps
야전 이동 취사반	mobile field mess
야전 전화	field telephone
야전공병	Field Engineer
야전군 전술작전본부	field army tactical operations center (FATOC)
야전변소	field latrine
야전삽	e-tool
야전선	field wire
야전용 무전기	field radio set
야전 포병	Field Artillery (FA)
약실	chamber
약정된	coordinated
약진하다	make a rush
양각대의탁엎드려쏴자세	bipod-supported prone position
양팔 간격, 우로 나란히!	At double interval, dress right, dress!
양호한 상태를 위지하다, ~를	keep in good shape
양호한 상태에 있다	be in proper condition
어깨 걸어 총!	Sling, arms!
어서!	Come on!
어쨌든 (anyway)	be that as it may
억류자	detainee
얻어내다	draw
엄격한	rigid
엄밀한	rigid
엄폐와 은폐	cover and concealment
엄호사격	covering fire
엄호하다	overwatch
엄호하에	covered by
여군	WAC (Women's Army Corps)
여러 발의 탄약	rounds of ammunition
역순	inverted order
역습	counterattack
역종	service status
연결	link-up
연결점	linkup point
연구하다	work up
연달은 고지들	chain of hills
연대	regiment
연대 책임	joint responsibility
연마된	honed
연막	smoke
연병장	drill ground; parade ground

연속적인 질문	a burst of questions
연신 또는 전환 (사격)	lifting or shifting
연합사(연합군 사령부)	CFC (Combined Forces Command)
열병 또는 분열식 장소	parade ground
열병, 사열	review, inspection of troops
열병장	reviewing ground
열중 쉬어!	Parade, rest!
영점사격하다	zero in
영점을 획득하다	zero in
영점조준하다	zero in
영창	lock up; detention barrack
예광탄	tracer ammunition
예규, 내규	SOP
예방정비 점검 및 수리	PMCS requirements
예비	reserve
예비(대)	alternate
예비대	reserve
예상전개선	probable line of deployment
예상접근로	likely avenues of approach
예속 혹은 배속된 공용화기	assigned or attached crew-served weapons
예포경례	cannon salute
예포대	salute battery
예행연습	rehearsal
오고 가는	to and from
오랜 시간에 거쳐 입증된	time proven
오발	accidental discharge
오성신호탄	star cluster
온통,(~에서 ~까지)	all the way
와해	disruption
와해된	disorganized
와해하다	disrupt
완만하게	sluggishly
완전 군장 얼차려	pack drill
완전명령	complete order
완전히	cold
완충기	buffer
왜, 못할 리가 있겠어	why not ~
외곽경계	perimeter security
외박증	pass
외선작전	operation on exterior lines
외주경계	security perimeter

외출 금지	confinement to barracks
외출증	pass
요도	sketch
요청 표적	proposed target
용이하게 하다	facilitate
우로 나란히!	Dress right, dress!
우로 어깨 총!	Right shoulder, arms!
우발상황	contingency
우선권/우선권내의 우선순위	priority/ precedence
우측분대	right flank squad
우향 우!	Right face!
운동장	playing field
울창한 수림	heavy woods
원기를 북돋우다	cheer up
원기를 회복하다,(병 후에)	be on one's feet
원반	discus
원자 (핵) 포병	Atomic Artillery
원형탄창	drum magazine
원형흑점표적	bull's eye target
위병근무(보초, 순찰 등의 근무)	guard duty
위병근무를 서다	be on guard duty
위병소	guardhouse
위병장교	officer of the guard
위생병	aid man; medic; first-aid
위장(하다)	camouflage
위장도색	pattern painting
위장망	camouflage net
위장망(올이 굵은 삼베로 만듦)	burlap
위치확인	navigation
위태롭게 하다	jeopardize
위험한 처지	tight situation
위협적인 표적	jeopardizing target
위협종합	threat integration
위협평가	threat evaluation
윗덮개	forearm
유격전	hit-and-run warfare
유무선 중계	wire and radio relay
유무선 통합	RWI (radio-wire integration)
유무선 통합 체제	RWI system
유사하다, ~와	resemble
유선	wire
유선 도청	wiretapping
유선 침묵	telephone silence
유속	water velocity
유언비어	false rumor; canard
유인효과	canalizing effect

유지작전	sustainment operations
유지하다	retain
유지휴대(장비 및 물자)	sustainment load
유탄(hand grenade 와 rifle grenade가 있음)	grenade
유탄발사기 사수	grenadier
유행성 감기 (influenza)	flu
유효사거리	effective range
육군 이동 외과 병원	MASH(mobile army surgical hospital)
육군 지휘 통제망	ACCNET(army command and control network)
육안 관측	visual observation
육체적 강인성	physical toughness
육체적인 훈련	physical exercise
육해공군 통합군	unified services
은닉	cache
은닉장소, (탄약, 식량 등의)	cache
은밀하게	with stealth
은폐	concealment
을지포커스 훈련	Ulchi-Focus Lens Exercise
음성문자	phonetic alphabet
음성으로 송신하다	I spell
음송전송(통신)	voice transmission
음성회로	voice circuit
음어	code word
응급 처치	emergency medical treatment
응집력	cohesion
의가사 징집연기	draft determent for family
의도	hardships
의도	intent
의무	Medical Corps
의무 중대	medical maintenance
의무대	medical company
의무병	aidman
의무병역	compulsory service
의무실	dispensary
의무중대	medical company
의장대	honor guard; ceremonial
의전 및 연락	protocol and liaison
의탁무릎쏴 자세	kneeling supported position
이동 통제점	movement control point
이동가능성	trafficability

이동소사	traversing and searching fire
이동야전취사반	mobile field mess
이동표적	transient targer; moving target
이동하다,(구간)	move by bounds
이병	private
이봐	say
이봐(병사들)	men
이상	over
이상무!	All clear!
이상할 것 없다	no wonder
이용하다, ~을	capitalize on
이적행위	profiting enemy
이전	NLT (no later than)
이탈	deviation
익숙하다, ~에	be accustomed to; be used to
인가회로(비문통화)	approved circuit
인계/인수	passage
인공 호흡	artificial respiration
인공물	cultural feature
인공지물(인공장애물과 구분됨), (장애물 중)	cultural obstacle
인명구조대원	lifesaver
인수하다	to take over
인수하다, ~을(를)	take over
인식 및 식별	recognition and identification
인식표시	recognition signal
인원점호	check roll call
인위적으로 조성된 제한된 시계	induced limited visibility
인접 배치하다	collocate
인접 부대	adjoining unit
인접부대	adjacent unit
인접부대장	adjacent commander
인집지역	immediate area
인지하다	be alerted for
일과시간	duty hours
일교차 (낮시간 범위)	diurnal range
일렬종대로	in a single file
일반 및 특별 보초수칙	general and special guard orders
일석점호	tattoo roll call
일시적인 전투 중지	lull
일을 떠 맡다	take over
일조점호	morning roll call
임검	surprise visit of inspection
임무	mission
임무 교대 하면서	alternation roles

임무 교대될 때까지	until relieved
임무를 수령하다	receive mission
임무분담을 하다, ~에	be detailed for
임무에 특정한	mission specific
임무형 보호태세 (MOPP)	mission-orinted protective posture
입닥쳐	Shut up.
입대선서	oath of enlistment
입대하다	to enlist
입영통지서, 영장	notice of enlistment; draft call
자격사격을 실시하다	qualify
자동소총	automatic rifle
자문관 자격으로	in an advisory capacity
자물쇠	safety lock
자물쇠 잠궈!	Lock your weapon!
자물쇠 잠그고 탄알 일발장전!	lock, one round!
자발적 철수	withdrawal not under pressure
자산	asset
자연적으로 형성된 이동이 용이한 통로	natural lines of drift
자연지물, (장애물 중)	natural obstacle
자원입대	voluntary enlistment
자주전투차량	self-propelled combat vehicle
자해	self-infliction
작동순환	cycle of operation
작전 지역	zone of action
작전개념	concept of operations
작전계획 또는 명령 양식	operations plan or order format
작전명령	OPORD (operations order)
작전지역	AO (area of operations)
작전지대	zone of operations
작전지원	operational assistance
잔류작전	stay-behind operations
잔류접척분견대	detachment left in contact (DLIC)
잘 지적했다 (You pointed it out well.)	Point well taken
잘 보다, ~를	command a good view of
잠복척후	sneak scut

잠정계획	tentative plan
잡아당기다, ~을(를)	pull ~ in
잡역	fatigue duty
잡음 방해	noise jamming
장갑 인원 수송차	APC(armored personnel carrier)
장갑차량	armored vehicles
장대높이뛰기	pole vault
장대높이뛰기하다	pole vault
장애물을 개척하다	breach the obstacle
장전 및 잠금	loading and locking
장전거리	arming range
장전된 탄약	cartridge supply
장전불능	failure to chamber
장전손잡이	charging handle
장전장애	loading jam
재송하라	say again
재송한다	I say again
재집결 지점	rallying point
재집결지	rally point
재편성하다	reorganize
저격수	sniper
저고도 방공	low altitude air defense
적 매복부대	enemy ambush force
적 포로	EPW(Enemy Prisoner of War)
적과 교전하다	engage the enemy
적군	enemy forces
적으로부터 이탈하다	disengage from the enemy
적을 격멸하다	finish the enemy
작을 고착시키다	fix the enemy
적응시키다	orient
적의 도청	enemy tapping
적의 사기를 뒤흔듬으로써	by shaking his morale
적절한	pertinent
적점령 지역	enemy-held area
적정을 탐색하다	to feel for the enemy
적지	hostile territory
적토부	fill
적합하다, ~에	be fit for; lend itself to
전 전역(全戰(域)에 걸친	theaterwide
전개하다	deploy
전과확대	exploitation
전달	routing
전령통신 취급소	messenger station
전방 진지	forward position

전방관측소 무전병	FO RATELO (forward observer RATELO)
전방관측자	FO (forward observer)
전방항공통제관	forward air controller(FAC)
전복	subversion
전사면	forward slope
전상자 후송	casualty evacuation
전술 항공 요청망	tactical air request net
전술도보행군	tactical road march
전술작전본부	tactical operations center(TOC)
전술작전예규	TACSOP(tactical standing operation procedures)
전술적 행동철자	TACSOP(tactical standing operation procedures)
전술통제방책	tactical control measure
전술항공 지원반	tactical air support element(TASE)
전술항공 통제반	tactical air control party (TACP)
전술항공 통제본부	tactical air control center(TACC)
전술항공 통제체제	tactical air control system(TACS)
전시표적	full figure target
전자 장비 관측	electronic observation
전장 후송	battlefield evacuation
전장생존	combat survival; battlefield survival
전장정보분석	intelligence prepration for battlefield
전장지역 평가	battlefield area evaluation
전쟁원칙	principles of war
전진로	route forward
전진축	axis of advance
전차기동륜	sprocket
전차는 12시 방향에 있다	Tanks are at your 12 o'clock.
전초	outpost
전투 결과보고	after-action report
전투 휴대량관리	load management
전투가늠자	battle sight

전투간 철수	fighting withdrawal
전투간 현지교체	felief in place
전투공병	Combat Engineer
전투근무지원	CSS(combat service support)
전투근무지원병과	combat service support arm
전투근무지원부대	combat service support element
전투력의 통합, 노력의 결집	unity of effort
전투병과	combat arms
전투사기	fighting morale
전투식량	rations
전투원	combatant
전투이탈	disengagement
전투정찰(대)	combat patrol
전투지경선	boundary
전투지역	zone of action
전투지원	combat support
전투지원 및 전투근무 지원 부대	CS and CSS units
전투지원 중대	combat support company
전투지원공병	Combat Support Engineer
전투지원병과	combat support arms
전투지원부대 과업	tasks to combat service support unit
전투진지	BT (battle position)
전투편성	organization for combat
전파방해	jamming
전혀 없는 접근, (방해 지장 등이)	clear approach
절름발이 오리	lame duck
점령	capture
점령시간	time of occupation
점표적	point target
점호	taps
점호대형	muster formation
점화시키다	ignite
접근로	avenue of approach
접근불능지역	inaccessible terrain
접근하다,(~에)	close(with)
접적이동	movement to contact
접촉선	line of contact (LC)
접촉을 이탈하다	break contact
접촉점	Contact Point
정규군	RA(regular army)
정돈된	shipshape

정돈함	locker
정렬	alignment
정리하다	work up
정면종사	frontal enfilade fire
정밀도하	deliberate (river) crossing
정밀공격	deliberate attack
정보	Military Intelligence Corps
정비	maintenance
정비를 위한(군사행위의) 일시정지	stand-down for maintenance
정상	peak
정식간격, 우로 나란히!	(normal interval), dress right, dress!
정신적 기민성	mental agility
정연한	shipshape
정위치!	Stand, fast!
정점	peak; summit
정조준하다	correction
정찰(대)	patrol
정찰대	patrol
정찰로	patrol route
정찰하다	reconnoiter
정확히 수신하였으며 지시대로 시행하겠다	wilco
정훈	TI & E (Troop Information and education Corps)
제1국민역	eligible conscription status
제1보충역	stand-by replacement
제2국민역	disqualified conscription status
제2보충역	retired replcement
제거하다	eliminate
제거하다	knock off; breach
제대	echelon
제독제	decontamination
제독키트	decontamination kit
제반 규칙	rules and regulations
제방	embankment
제식훈련,(이동간)	movement
제식훈련,(정지간)	formation
제압사격	suppressive fire
제외하다	preclude
제외하다, ~을 ~에서	leave something you of
제자리 뛰어갓!	In place, double time, march!
제자리 섯!	Halt!

제한된 도로망	limited road net
제한사항	restriction
제한하다	mask
조공	supporting attack
조달하다	procure
조립병	encoder
조명	illumination
조식은 T-ration, 중식은 MRE, 석식은 T-ration으로 한다	T-MRE-T
조정 불량	maladjustment
조준선 정렬 및 정조준	sighting and aiming
조준선 정렬을 하다	align the sights
좀 더 접근하라.	Bring ~ in a little!
좁은 간격, 우로 나란히!	At colse interval, dress right, dress!
종격실	corridor
종대	column
종대(횡대)대형	column(line) formation
종대대형	column formation
종사	enfilade fire
좋다,(몸 상태가)	feel fit
좋아하다, ~를(을)	be fond of
좌로-다시	Left-As you were!
좌선 사격끝	Clear on the left!
좌측분대	left flank squad
좌표	coordinates
주공	main attack, main effort
주도권	initiative
주력	main effort
주유	lubrication
주전투진지	primary fighting positions
주파수	radio frequency
주한 미군 의무 사령부	MEDCON-K (Medical Command in Korea)
주한미군 전술표적 목록	Armed Forces Korea Tactical
준비단계	preparation phase
준비된 사수로부터 사격개시!	Commence firing when ready!
준비명령	warning order
준비명령을 하달하다	issue a warning order
준수하다	observe
준칙	tenet
줄줄이 좌로 갓!(이열종대로)	Column of twos to the left, march!
중(中)대전차화기	MAW(medium antitank weapon)
중/소위	lieutenant

중간 목표	intermediate objective
중고도 방공	medium altitude air defense
중대 치중대	company train
중대장	company commander
중령	lieutenant colonel
중사(육군)	SFC(sergeant first class)
중심	nucleus
중심(표적의)	bull's eye
중앙분대	center squad
중요 지형지물	key terrain
중요지형	key terrain; ground of vital importance
중요지형지물	key terrain feature
중요한 지형	key terrain
중화기중대	heavy weapons company
증강된	augmented
지근지역	immediate area
지급하다	issue
지대공 방어	surface-to-air defense
지도와 나침반에 의한 정치(定置)	orientation by map and compass
지도정찰(지상 또는 항공정찰의 예비단계	map reconnaissance
지상관제 요격	ground controlled interception(GCI)
지상 대기	ground alert
지상 전방항공 통제관	ground forward air controller (G-FAC)
지상관측	ground observation
지상전술계획	ground tactical plan
지선	tangle foot
지시	instruction
지역수색	area search
지역표적	area target
지연(작전)부대	delaying force
지원병역제도	voluntary service system
지원사격	supporting fire
지쳐버리다	be exhausted
지표(물)	landmark
지하저항운동	underground resistance
지향 사격자세	crouching position
지형격실	compartment of terrain
지형모형	terrain model

지형물	geographic feature
지형부호	topographic symbol
지형분석	terrain analysis
지형적 정상	topographic crest
지형지물	terrain feature
지형판단	estimate of terrain
지형평가	terrain evaluation
지형평가 5개 요소	five military aspects of terrain, OCOKA
지휘 및 통신	command and signal
지휘 및 통제	command and control
지휘관	CO (commanding officer)
지휘권 남용	misuse of leadership
지휘권승계	succession of command
지휘소	command post
지휘자에게 요구되는	required of leaders
지휘체계	chain of command
지휘하에, ~에	under the command of
직권 남용	misuse of authority
직사각형	rectangle
직속상관 등을 칭하는 속어	old man
직접항공지원본부	direct air support center(DASC)
진단서	medical certificate
진열검열	showdown inspection
진지강화	consolidation
진지강화하다	consolidate
진지구축자재	construction material
질리다	be fed up
질병 증명서	certificate of disability
짊어지다(등위나 어깨에 매다)	hump
집결지	AA(assembly area)
집단 반항	mutiny
집중하다	mass
집총경례	rifle salute
집합(대형)	formation
집합나팔	muster call
집합하다	fall in
집행	enforcement
징계 위원회	disciplinary committee
징계, 견책	official reprimand

징병검사	physical examination for draft
징병소	recruiting station; induction station
징병제도	conscription system
징집기피자	slacker
징집면제	draft exemption
징집연기	determent of enlistment
징입연도	conscription year
징집영장	draft notice
징입영장 (draft notice)	draft call
징입유예	temporary draft exemption
짧은 휴식	break
쭉	all the way
찔러!	Thrust and hold, move!
찔러동작	thrust, passado
찢어버리다	tear up
차개	ejector
차량화 소총소대	MRP(motorized rifle platoon)
차려총!	Attack position, move!
차려총자세	attack position
차렷 자세를 하다	stand at attention
차렷!	Attention!

차례	turn
차상급부대	next higher command
차장(은폐)	concealment
차장하다 (아군의 작전을 적군이 알지 못하도록 하게 함)	screen
차폐	defilade
차폐사격진지	defilade firing position
착륙지대	landing zone (LZ)
참을 수 없다, ~을	can't stand
참조점	reference point
참호	trench
참호 속에 배치되다	be dug in
참호선	trench line
참호진지를 구축하여 방어 편성을 하다	be entrenched
창백한	pale
찾아내다	locate
책임감	sense of responsibility
책임구역	sectors of responsibility
책임을 명시하다	delineate responsibility
책임지다	attend

책임지역	area of responsibility
척후	scout
척후기(정찰용 비행기)	scout
척후병	scout
천천히 송신하라	speak slower
철갑탄	armor-piercing ammunition
철수	withdrawal
철수시간	time of withdrawal
철조망	wire
철퇴	retirement
첨병	point man
청소하다	police
체력단련	PT(physical training), physical exercise
체력단련훈련	physical training
체온을 대다, ~의	take one's temperature
체육관	gym (gymnasium)
체육훈련	PT (physical training)
체조 대형으로 벌려	PT formation, move!
체형	corporal punishment
초 교파적 예배	nondenomination al service
초등군사반	Officers' Basic Course(OBC)
총가	rifle stand
총강	bore
총걸이 고리	stacking swivel
총검	bayonet
총검술 연습	bayonet exercise
총검술 훈련	bayonet drill; bayonet training
총검술대형으로 벌려	form for fencing gear
총검술용 방호구	muzzle
총구	Bore clear!
총구 이상무	small of stock
총목	receiver group
총몸뭉치	receiver group
총상	bullet wound
총성	shot
총열	barrel
총열뭉치	barrel assembly
최기 출발시간	earliest time of move; earliest start time; EST
최대로 이용하다, ~을	take advantage of
최저표척사(격)	grazing fire
최종목표	final objective
최종협조선	final coordination line

최초계획을 수립하다	make a tentative plan
최초공격대형	initial attack formation
최초목표	initial objective
최후 협조선 (폭격 개시선)	final coordination line
최후방어사격과 결합 되면	tied in with an FPF
추진장약	propellant
축성	fortification
축어적(逐語的)인	verbatim
출격(전투기의)	sortie
출두하다, ~에	report to
출발하다	set out
충격 효과	shock effect
취급하다	handle
취사 근무	cookhouse fatigue
취사근무	KP (kitchen police)
취사담당관	mess sergeant
취사병	KP (kitchen police)
취약점을 찾아내다	locate a weakness
취약한 측방	vulnerable flank
취침나팔	taps
취침점호	bed-check
취침하다	go to bed
측면으로	laterally
측방기동	lateral maneuver
측사	flanking fire
측후소	meteorological station
치료소	clearing station
치명적 효과	deadly effect
친구	buddy
침구정돈을 하다	make one's bed
침대	bunk
침투	infiltration
타박상	bruise
탄주; 탄환	projectile
탄띠	cartridge belt, pistol belt
탄알집	magazine; case
탄알집 끼어!	Insert magazine!
탄약	cartridge
탄약병(彈藥兵)	ammunition detail
탄약수	ammunition bearer
탄착군	shot group
탄착점을 예언하다	call your shot; call the shot
탄창	magazine
탄창식	magazine-fed
탄피	cartridge case
탄환	bullet
탄흔	bullet-mark
탈영	desertion

탈취	seizure
탈취 및 확보	seizing and securing
탈취 및 확보하다, ~을	seize or secure
탈취하다	to seize
탑승병력	mounted forces
탑재지대	pickup zone (PZ)
탕탕하는 총소리	cracking of shots
태업	sabotage
통과지점	passage point
통신	Signal Corps
통신 내규	communication SOP(standing operation procedure)
통신두절	loss of communication
통신 보안	COMSEC (communication security)
통신본부, 통신취급소	COMMCEN(communications center)
통신분견대	signal party
통신운용지시	SOI (signal operation instructions)
통신준칙	SSI (standing signal instructions)
통신파견대	signal detachment
통신 피지원 부대	signal supported unit
통신 훈련	COMEX(communications exercise)
통신병	RATELO (radio telephone operator), signal man
통신보안	COMSEC
통신보안규정	COMSEC REGULATION
통신운용지시	SOI (signal operation instruction)
통신전자 보안지침	communications and electronics security instructions
통신전자운용지시	CEOI (communication electronics operations instruction)
통제방책	control measure
통제선	Phase Line (PL)
통제선(Phase Line)	PL
통합부대	cohesive element
통합하다	integrate
통합하다(시간 상으로)	synchronize

통합하다; 모아합치다 (장비 등을)	consolidate
통합화력지원반	FIST(fire integrated support team)
통화 약어	procedure words (prowords)
투명도	overlay
투창	javelin
특공소총소대	ranger rifle platoon
특별작업	special detail
특수목적공격	special-purpose attack
특수임무	special detail
특수작전	special operations
틀림없이 그렇게 할께!	You bet!
팀스피리트 훈련	Team Spirit Exercise
파면	dismissal
팔굽혀 펴기를 10회 하다(do 10 pushups)	drop and give 10
패배하다	be beaten
편조, 부대편성	task organization
폐쇄등고선	closed contour line
포대	BTR; battery
포대장	battery commander
포병	Artillery
포병인	cannon-cocker
포병포대	battery
포복	crawling
폭격 금지선	No bomb line (NBL)
폭격선	bomb line (BL)
폭격안전선	bomb safety line (BSL)
폭동데모	riot demonstration
폭격	violence
폭약	demolitions
포가	demolition
표적	target
표적 좌표	tatget coordinates
표적공격시간	time on target (TOT)
표적에 명중시키다	hit the mark
표적우선순위	priority of targets
풀다,(자물쇠를)	unlock
풀어 총!	Take, arms!; Unpile, arms!
피한 길이 없다	It can't be helped.
필수 과업	essential task
필수 임무과업	mission-essential task
하고 있는 행동을 멈추다	knock it off

하극상	usurpation of power by the lower to the high
하기식 집합	retreat formation
하달	issuance
하사(육군, 해병대), 공군병장	SSG (staff sergeant)
하사, 중사, 상사 등 부사관을 통틀어 호칭함	Sergeant
하역작업	off-loading
하위 집단	subdivision
하천선 작전	river line operation
학생 징집연기	draft determent for education
한국군	ROK Armed Forces
한미 군사위원회 회의	Military Committee Meeting(MCM)
한미 안보협의 회의	Security Consultative Meeting(SCM)
한미 연합군 사령부	ROK-US Combined Forces Command
한번에 하나씩	one at a time
한창 친행중이다	be in full swing
할 수 있다, ~을	be capable of
할당(배분)하다	allocate
할당구역	assigned sector
함몰도로	sunken roads
합동 공중 공격반	Joint Air Attack Team(JAAT)
합동 및 연합작전	combined and joint operations
합동공지작전 체제	joint air-ground operations system (FAGOS)
합동작전 본부	joint operations center(JOC)
항공	Aviation corps
항공 의무 후송	aeromedical evacuation
항공기의 공격방향	attack heading
항공차단	air interdiction
항명	insubordination
항복, 귀순	submission
해두보 (상륙작전시)	beachhead
해병 기동부대	marine task force
해산하다	dismiss
해상 박명종	EENT(end of evening nautical twilight)
해상 박명초	BMNT(beginning of morning nautical twilight)
해안 장애물	beach obstacle
해역병	decoder

핵 및 비핵 화력	nuclear and nonnuclear fire
핵심	nucleus
핵안전선	nuclear safety line(NSL)
행사하다	exert
행정 및 군수지원	administrative and/or logistic support
행정지원병과	Administrative Support Branches
향토예비군	HDRE(Homeland Defense Reserve Force)
헌병	Military Police Corps
헌병감	Provost Marshal General
험란한 지형	difficult terrain
헤어나지 못하다	be stuck with
헤쳐!	Fall out!
헬기장	heliport
헬리콥터 발착장	heliport
현역	active service
현재 보유하고 있다	be on hand
현저한	marked
협조	coordination
협차사격	bracket fire; straddling
형판	template
호출부호	call sign
혹	lump
혼선	cross talking
홀수와 짝수	odd and even numbers
화기분대	weapons squad
화기훈련	weapons training
화력	fire power
화력의 우선권	priority of fires
화력지원	fire support
화력지원반	fire support team (FIST)
화력지원장교	FSO(fire support officer)
화력지원팀	FIST(fire support team)
화력지원협조본부	fire support coordination center
화력팀(부대는 2개 fire team으로 구성됨)	fire team
화생방	CBR(chemical, biological, and radiological)
화학	Chemical Corps
화학, 방사능, 연막 및 화염작전	chemical, radiological, smoke and flame operation
화학작용제	chemical agent

확보하다	retain; secure
확실한 장악능력	firm grasp
확인점	check point (CP)
확인하다	I verify
환자 집합	sick call
환자집합하다	go on sick call
홱 몸을 굽히다 (맞지 않으려고)	duck down
횡격실	cross compartment
횡대	line
횡대대형	line formation
횡사	traversing fire
후미분대	trail squad
후방지역경계	rear area security
후방지역작전	rear area operations
후속부대	follow-on forces
후송 계통	chain of evacuation
후송하다,(급히)	sped to the rear
후임자	replacement
후퇴작전	retrograde operations
훈계	admonition
훈련병	trainee
휴대용 장비 (탄약, 무기 등)	load-bearing equipment(LBE)
휴게(오락)실	dayroom
휴대용 대전차 공격 미사일(M47)	Dragon
휴대용 폭약 및 기폭장치	satchel charge
휴식계획	rest plan
흑점	bull's eye
히죽이 웃다	grin

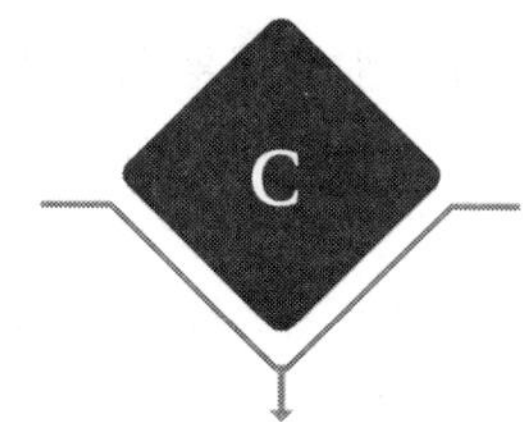

Special Expressions

특수표현

about face	결정, 지시 등을 바꾸다. (To reverse a decision) The Captain had to make and about face in orders because of the sudden enemy attack.
ace in the hole	불의의 수단, 무기, 상황. (A surprise; a surprise situation) The reserve force is the general's ace in the hole.
all for it	완전히 동의하다. (Agree entirely) This is a good plan and I'm all for it.
all out, go	모든 수단을 동원하다. (Try every means available) It is necessary to go all out to capture this enemy position.
attack jumps off	공격을 개시하다. (Attack begins) The attack jumps off at 04:30 hours.
attack stalls	공격이 지연되다. (Attack is delayed) Our attack stalls whenever we encounter heavy fire.
AWOL (Absent without leave)	무단이탈, 군무이탈. He was AWOL for four days; he will be court-martialed.
baby	문제, 과제, 임무. (Problem) Making the estimate of the situation is your baby.
backbreaker	어려운 일, 고되고 힘든 과제 (Difficult job) This tool turn-in is a backbreaker.
backed up	보강하다, 강화하다. (Reinforced) The 25th Infantry will be backed up by supporting artillery.

bad shape	어려운 상황, 형편, 나쁜 육체적 상태.(In difficult position; in bad physical condition) Our reserve failed, so now we are really in bad shape.
balled up	혼란스러워지다, 바르지 못하다. (Confused; wrong) When the transportation didn't arrive, things became balled up.
bark up the wrong tree	엉뚱한 곳에서 답을 찾다. (Look for a solution in the wrong place) Smith couldn't get a correct answer because he was barking up the wrong tree
bawl out	심하게 꾸중하다, 야단치다. (Reprimand; speak harshly to) It is not always wise to bawl out subordinates for every little mistake.
best around the bush	본론을 말하지 않고 말을 돌려서 하다. (Delay coming directly to the point) Instead of telling me the truth, he tried to beat around the bush.
best me	전혀 모르겠다. (don't understand this; doesn't make sense to me) What does this dispatch mean? It beats me.
beef up	강화, 증강하다 (reinforce) We'll have to beef up the main effort with extra artillery support.
be up against	맞다, 대치하다 (to be faced with) He really doesn't know what he is up against when he accepts a mission like that.
bend over backward	열성적으로 위하다, 애쓰다 (to try hard; do all in your power) Bill bends over backward to be nice to everyone.
big deal	대수, 큰일 (anything important, satisfying, or lavish) Winning the battle was a big deal.
big picture	전체적인 안목 (complete view or plan; overall view) After the briefing, he could see the big picture much better.

bird's-eye view	전체적인 설명, 해설 (overall explanation) Before staring the movement, I'll give you a bird's-eye view fo the plan.
black out	정전, 등화관제 (without light or security) All vehicles will be blacked out tonight.
blast	강타하다 (strike a hard blow) This tank is able to blast any pillbox.
Blitz	전격전, 기습(하다) (attack and overcome quickly) With this plan, we hope to blitz the enemy.
blow up, blow your cool	top, blow your stack; blow a gasket; blow a fuse 화내다 (to become angry or excited; to lose self-control) The colonel blew his when he saw the report.
bog down	지연하다, 중지하다 (to slow up or stop) The attack will bog down if you do not get us more fire support.
boil down	간단하게 정리하다 (to reduce to basic form) Please help me boil down this question to a form in which it can be answered.
brass, the	고위급 장교들 (high-ranking officers) The brass will inspect us today, so get everything ready.
breakdown	실패, 고장, 두절 (failure) The message didn't get through because of communications breakdown.
break out again	다시 생기다, 일어나다 (occur again) We believe heavy firing is about to break out again.
bring out a point	요점을 제시하다 (illustrate) I want to bring out the following point as a summary.
bring to light	밝혀지다, 알게 되다 (reveal; illustrate) This intelligence is likely to bring to the light the presence of new enemy reinforcement.
broad picture	상화에 대한 대략적인 그림, 내용. (a general view of a situation) The VI corps gave us the broad picture of corps operations.

buddy seat	운전사 옆 좌석 (seat for assistant driver) I'm wondering if the person seated in the buddy seat can turn on the electricity.
bulldog tank	M/MA 탱크 It is easy to identify the bulldog tank.
bump into	우연히 만나다, 마주치다 (meet or engage unexpectedly) Opposing flanking movement tend to bump into each other.
button, on the	어김없이, 정확히 (exactly correct) He answered every question on the button; his grade is 100 percents.
button up	종결 짓다, 끝내다 (finish; close) This memorandum will serve to button up the issue.
call your shot	정확히 예측하다 (판단하다) (estimate correctly) The enemy is trying to surround us, at 07:30; you certainly know how to call your shot.
can make it	제시간에 도착하다 (arrive on time; be there) I can make it to the post by.
canned message	준비된 메시지 (a prepared message) During practical exercises, we used many canned messages.
can't see the forest for the trees	세부적인 것에 집중한 나머지 전체적인 것을 못 보다 (too busy with details so see the overall plan) All the men in the division are not kept informed; therefore, they can't see the forest for the trees.
carry on with	계속 하다 (continue) After the briefing, each section will be able to carry on with its work.
carry out an order	명령을 수행하다 (do or execute an order) To carry out an order efficiently, one must first understand it.
check!	맞아! 맞다! (right! correct!) Is this what you wanted? Check!

check up on	조사하다, 살피다, 확인하다 (verify; inspect) Good staff officers cheek up on orders being carried out correctly.
chewed to pieces	심하게 고장 나다, 박살나다 (severely damaged) The enemy forces were chewed to pieces by our accurate artillery fire.
chicken out	회피하다. 도망하다 (to lose heart; abandon a course of action) The attack was proceeding nicely until the second platoon chickened out.
chiefs and indians	지휘자와 부하들 (leaders and assistants) There are too many chiefs and not enough indians in this office.
chopper	헬기 (helicopter) We had our first ride in the new chopper.
clear cut	확실한, 명확한 (definite; specific) It was undoubtedly a clear-cut victory.
clear up	설명하다, 해결하다 (explain; solve) I don't understand this point; would you care to clear up it up for me?
click	일심동체로 일하다 (work harmoniously) Well-drilled units can be expected to click in any situation.
close shave	겨우 성공하다, 겨우 사고를 면하다 (barely successful; barely avoiding an accident) The enemy attack failed to destroy the bridge, but it was a close shave
cloud the issue	어렵게, 복잡하게 만들다 (make something less easy to understand) His lengthy explanations tend to cloud the issue.
cock-and-bull story	믿을 수 없는 내용 (an unbelievable story) His explanation was just a cock-and-bull-story

comp through	세밀히 검토하다, 수색하다 (Search thoroughly) Someone ought to comb through these carefully to fine any errors.
come out in good shape	공격에 잘 대응하다 (withstand the attack) If all regiments come out in good shape, the plan will have been successful.
cook up	준비하다, 계획하다 (prepare; make a plan) Try to cook up something new to surprise the enemy.
crack down	심하게 대하다 (Deal with severely) He will crack down on us if we don't complete all of the reading assignment.
crackup	사고 (Accident) Three fighter planes had a crackup on leading.
crank up	시작하다, 게시하다 (start) It's nearly time to cross the initial point; crank up your motors.
crop up	갑자기 나타나다, 돌발하다 (Appear suddenly; happen) Remember how you handled the situation; it is likely to crop up again.
curtain of fire	심한 총격이 벌어지는 지역 (Barrage; concentrated artillery shellfire) Our troops advanced behind a heavy curtain of fire.
cut off	포위되다, 고립되다 (Surrounded; isolated) In the defense of an extended front, some units may be cut off for a while.
cut to the bone	최소한으로 줄이다. (Reduce to absolute minimum) In the defense of an extended front, some units may be cut to the bone.
dark, in the	아무것도 모르는 상태 (Ignorant of) What does this mean, G-2? Don't keep me in the dark.
dead wood	쓸모없는 (Worthless) Can I transfer him? He's just deadwood in the section.
deal me in	나도 포함시켜 달라 (Include) Deal me in if you go the movies.

deal me out	나는 포함시키지 말라 (Exclude; do not rely or depend on) Deal me out if you go to the movies.
deuce and a half	2 1/2톤 트럭 (ton truck) That deuce and a half has plenty of power.
dig out	발견하다, 공부, 연구하다 (Discover; study) You'll have to dig out these experience factors from FM 101-10.
done for	끝나다, 패하다 (Finished; defeated) If the storm hits us now, we'll be done for.
done in	피곤하다 (Tried) After sitting in class all day, I'm really done in.
Don't bet on it.	보장 못 한다, 의존하지 말라 (Don't depend or relay on it) I will try to be there but don't bet on it.
Don't fight the problem	싸우려 들지 말라; 대응하지 말고 그냥 받아드려라 (Accept this as they are; don't offer resistance) I know that this is a difficult matter, but don't fight the problem; work slowly and carefully.
dope, bone-head, knot-head, knuckle-head	느리거나 멍청한 사람(A slow or stupid person) Only knot-head like Tom would do that!
dope	정보, 마약 (Information, especially from a reliable source) The G-2 gave us the latest dope in the enemy situation.
doughboy	보병 (Infantryman) In the final analysis, it's the doughboy who has to take the objective.
down a blind alley	막다른 골목에 도달하다 (Reach an impasse) You're down a blind alley, Bill; we won't get extra trucks for this movement.
down the home stretch	끝내다, 마무리하다 (Tank transporter) In weeks, we'll be going down the home stretch in phase I Operation.
dragon wagon	탱크 수송차량, 수송기 (Tank transporter) If we didn't hae the dragon wagon, I don't know how we could move that tank.

draw the line	선을 긋다, 그이상은 절대 못하게 한다 (Prohibit from going any further) When it comes to smoking during instruction, he draws the line.
dropped the ball	실수하다, 기회를 놓치다 (Committed an error; missed an opportunity) We certainly dropped the ball when we didn't follow the instructions in the manual.
dud, a	실망스런 실패작, 불발탄(A disappointing failure) The comedian was a dud.
eager beaver	무엇이든 열심히 앞장서서 하는 사람 (First in line; volunteer) Captain Bean is an eager beaver.
eggs in one basket, all your	너의 모든 것을 한곳에 집중시키다 (Put all your resources in one effort) Make three separate plans; don't put all your eggs in one basket.
eyeball to eyeball	아주 가까운 거리에서 (At bayonet's distance; face to face) Our doughboy fought eyeball to eyeball in the enemy trenches.
eye on, keep an	감시하다 (Watch closely) You ought to keep an eye on enemy activity.
fan out	흩어지다, 펴지다 (Disperse; deploy) The leading battalion will fan out to cover our advance.
feed	보내다, 송신하다 (send; transmit) The job of divisions to feed G-2 with information on daylight movements of the enemy.
feet on the ground	침착하고 논리적인 (sensible; logical in manner) Out platoon leader is one who keeps his feet on the ground.
fifty-fifty	반 반, 대 (half and half; equal) Let's split the dinner bill fifty-fifty
fill the gap	뚫리는 것을 방지하다, 돌파를 막다 (stop a breakthrough) Our reserve arrived in time to fill the gap.

fill in the gaps	모르는 것을 보충하다 (provide missing information) We filled in the gaps on our intelligence of the enemy by careful observation
fingertips, have at your	가까운 곳에 준비하다, 대기하다 (have available, ready) What's your estimate, G-3? You ought to have it at your fingertips.
finish off	끝내다, 섬멸하다 (to defeat or and) When that last strongpoint surrenders, we can finish off the rest easily.
firm up	계획 등을 미치다, 끝내다 (complete; finish) The general wants us to firm up all plans by 13:00 hours.
fix the enemy	적을 무마시키다 (keep the enemy inactive) The base of fire serves to fix the enemy in place.
flap, in a	큰 혼란, 혼돈에 빠져있는 (in great confusion) For lack of information, the staff was in a flap.
flip, flip your lid	화내다, 흥분하다, 자제력을 잃다 (to become angry or excited; to lose self-control) The colonel flipped his lid when he was the report.
follow-up	감독하다, 지켜보다 (verify; supervise) Staff officers are to follow-up on all plans and orders issued.
for the birds	믿을 수 없는(unrealistic) The answer he gave us is for the birds.
freeze	부동, 멈춤 (restrict; stop in place) If there's no time to evacuate civilians, it will be necessary to freeze them in place.
fresh blood	지원군, 증강된 군인원 (reinforcements or replacements) The enemy was waiting for fresh blood before making a new attack.
gadget	기계적인 물건 (any mechanical thing) What's that gander, Kick? It's a special tool for radios, John.

gear	장비(equipment) All right, men make sure all your gear is clean.
geared to	맞추어서, 맞게 (adjusted to) Our advance must be geared to that of the adjacent divisions.
get ahead	앞서가다, 추월하다 (advanced) Don't try to get ahead too fast.
get a kick out of	재미있어 하다, 즐기다 (to enjoy) I really get a kick out of that person.
get a lucky break	운이 좋다. 좋은 기회가 주어지다 (be fortunate; gain an advantage) We'll get a lucky break if there's no heavy participation.
get around a difficulty	해답을 찾다 (find an answer) A resourceful staff officer knows how get around a difficult.
get away with it:	요행이 성공하다 (succeed by taking a chance) You may get away with it this time, but you won't always be so lucky.
get by with	해내다 (succeed; mange) We may have to get by with only a few men.
get in the act	허락이나 동의 없이 참여하다 (join on doing something, usually without being asked) Every time we plan to go to town, he tries to get in the act.
get in touch with	연락이 닿다 (communicate with) If you want to get in touch with him, call him at his home.
get in the stick	준비하다 (get ready) We have to get on the stick if we want to get to the next village by nightfall.
get on with	계속하다. (continue) Excuse the delay, gentlemen; let's get on with the briefing

get over	(어려움을) 극복하다 (to overcome the bad effects) He got over his illness quickly.
getting the hang of	대체적으로 이해하다 (to understand the general idea) I can't get the hang of the procedure.
get the picture	상황을 파악, 이해하다 (understand the situation) After the explanation, did you get the picture?
get the word	정보를 알아내다, 듣다 (to find out the information) I got the word about the next problem.
get together with	누구와 만나다, 모이다. (meet, confer) G-1 and G-2 get together with each other on all operations.
get your hands on	손에 넣다, 구하다 (obtain) Do you know where you can get your hands on any more ammunition right away?
gimmick	편법적 수단, 도구 (an unorthodox method to device) This gimmick is causing the bearing to wear away.
gingerly	조심스럽게, 부드럽게 (softly; carefully) Mines must be place in position gingerly.
gismo; thingamajig	명칭을 모르는 사물을 가르키는 단어 (anything whose name you don't know) What's that gismo? it's a new safety device.
give a rundown on	정리하다, 요약하다 (give a summary of) Give me a rundown on the situation.
give away (a giveaway)	탄로 나게 하다, 누설하다 (fail to deceive (failure in deception)) Those newly cut trees in that area sure give away his position.
give him the green light	동의하다, 지속할 수 있도록 허락하다. (give your agreement (signal to continue)) The bottleneck is cleared up, so give him the green light.
give him the works	벌주다 (punish) Now is the time, sir, to give him the works.

give it the once-over	조사하다, 검색하다 (scrutinize; examine) Here's the order, Ed; give it the once-over, will you?
give it the shotgun method; shot gun it	짐작으로 일을 서두르다 (do something quickly, usually by haphazard guessing) Now don't try to give it the shotgun method; I want you to be specific.
give the nod to	승인하다 (approve) Tomorrow he may decide to give the nod to your suggestion.
gobbledygook	이해할 수 없는 정보 (unintelligible information) What's that gobbledygook? It's a cryptographic message; here's the clear text
go for broke	마지막 필사적이 사도 (A final desperate effort) We will have to go for broke to face certain defeat.
go for record	공시적으로 무엇을 진행하다 (Do something officially) The general remarks to his staff just before entering combat for the first time "now we go for record"
go though the mill	어려운 경험을 하다 (Experience; undergo hardship) The man have to though the mill to be successful.
go to pot	망가지다, 무용지물이 되다 (Deteriorate; fall to pieces; become worthless) Without good leadership, any unit will soon go to pot.
going off halfcocked	준비 없이 말하거나 행동하다 (Speaking or acting without proper preparation) That's what we get for going off halfcocked.
goldbrick	게으른 사람(A lazy person) Keep your eye on him; he's a goldbrick.
good deal, it's a	좋은 상황이나 일을 뜻하는 감탄사, 표현 (Wonderful; to one's liking) How do you like your new job? It's a good deal.
goof off	일을 피하다, 비보 같은 행동을 하다. (To avoid work; to act foolishly) That young soldier goofs off all day.

goose eggs	숫자, 타원형 (Zero or elliptical in shape) He received a goose egg for his effort in the test.
graveyard shift, the~	근무, 야간작업 (From to; night work) It's your turn as your duty officer next month for the graveyard shift
grease monkey	기능공 (Mechanic) It is easier for a grease monkey to repair the truck.
grounded	이륙, 운전 승인이 안 나다 (Not permitted to fly, or drive) All our preplanned observation missions were grounded during the snowstorm.
ground rule	정상적인 절차, 정해진 기복적인 규율 (A normal procedure; a fixed rule) There's a ground rule here against such actions.
ground work	기본적이고 세부적인 활동, 작업 (Basic detail work) The early ground work on the plan was performed by G-3.
guts	용기, 담 (The will to fight; bravery) The doughboy with the rifle is the one who has the guts in battle.
hits the nail on the head	딱 맞다, 정확하다 (is exactly correct) Your solution hits the nail on the head.
hit the road	출발하다; 시작하다 (begin; stary) Well. let's hit the road; we still have hours to return to the command post.
hit the sack	잠자러 가다 (go to sleep) I'm tired; I think I'll hit the sack.
hold down	최대한 줄이다 (reduce to a minimum) To preserve surprise, we will have to hold down all noise.
hold out	계속 저항하다 (continue to resist) You must hold out until relieved by the linkup forces.
hold up his end	맡은 임무를 수행하다 (do his share of the work) Each staff officer must hold up his end of the load.

hookup	통신 연결 (join in communications; juncture) Can you contact the battalion commander by some hookup?
hook up with	연결하다 (join, as in communications) Out flank patrol is to be hook up with those of the adjacent.
horns of a dilemma, on the	궁지에 처하다(in a difficult position) What's your opinion? I'm on the horns of a dilemma.
horse around	놀다 (engage in rough boisterous fun) His chances for promotion are poor, for he horses around instead of working.
hot potato	어려운 문제, 상황 (difficult problem; a situation) That question is some hot potato; have you any ideas on it?
hot shot	뛰어난 인물 (very efficient; very important person) This regimental S-3 is really a hot shot on everything he does.
hot spot	궁지 (a predicament; a troublesome situation) The company, sup off from the battalion by the enemy troops, was in a hot spot.
hundred eighty degrees you of phase	완전히 틀린 (completely wrong) I was a hundred eighty degrees out of phase on that last exam question.
I'll buy that	동의하다 (I agree) G-4, your logistic plan looks OK to me; I'll buy that.
I'll go along with that	동의하다, 수용하다 (I agree or accept) Captain, your plan looks OK to me ; I'll so along with that.
I'll settle for that	(좀 미흡하지만) 수용하다(I'll accept that) I need trucks but have received only , so I'll settle for that.
I'm sold on	... 에 반하다. (I like) I'm sold on the idea of small unit rotation.

in a sweat	초조해 하다, (to be in a difficylt situation) He was in a sweat about the general's inspection.
Indian	부하 (a subordinate) Well, here are my orders; I'll just be an indian in the Pentagon.
info	정보 (축약형) (shortened form of information) The G-2 will provide the info on intelligence matters.
in the air	미결상태의, 불확실한 (undecided; uncertain) All plans are in the air since the old man received his orders.
in tune with	합치하다 (agree; in accord with) All division operations must be in tune with the general's concepts.
ironclad rule	불변의 원칙 (invariable; without exception) It is an ironclad rule that subordinates show respect to supervisors.
iron out	세부적으로 조정하다. (arrange the details of; coordinate) You must see the division adjutant to iron out this respect to program.
It's beyond me.	나로서는 이해가 안 돼. (I don't understand this) I can't understand why there's no news; it's beyond me.
It strikes me	생각이 들다. (I think; believe) It strikes me that we need reinforcements just to be safe.
jack of all trades	다재다능한 사람, 재주꾼, 만능 (a versatile person) That boy is a jack of all trades in sports.
jigsaw	수수께끼 (riddle; all mixed up) G is the one who most solve the jigsaw puzzle of enemy strategy.
jump ahead of	한걸음 앞서다(to do anything prematurely) I'm one jump ahead of you; I've already arranged that.
jump off (of an attack)	개시하다 (begin; start) Out attack is to jump off tomorrow morning.

jump the gum	성급하게 행동하다 (to do anything prematurely) Smith is nervous about the project, and he is likely to jump the gun.
jump to conclusion	속단하다, 성급하게 단정하다 (decide too quickly) This is serious; don't jump to conclusions.
keep an eye on(or out)	지켜보다, 감시하다 (to watch or watch for) Keep an eye out for Jane. She will be here soon
keep clear of (or steer clear of)	피하다(avoid) Warn your men to keep clear of that minefield near you.
keep close tabs on	주시하다 (watch closely) We will have to keep close tabs on that enemy armored division.
keep hammering at	계속 타격을 가하다, 강조하다 (continue to attack; emphasize repeatedly) Keep hammering at them in that area until we open their line.
keep the enemy off your back	적이 방해하지 못하도록 방지하다 (prevent the enemy from bothering you) In a with drawl, the purpose of the detachment left in contact is to keep the enemy off your back.
keep your finger on	통제하다 (control or keep in contact with) You have th keep your finger on everyone involved at all times.
keep your finger crossed	행운을 빌다 (hope for the best) We expect to win, but the enemy is resourceful; keep your fingers crossed.
deep your wits about you	침착하게 행동하다 (be calm) Don't get excited during the enemy attack; just keep your wits about you.
keystone	중점 (the principal point) The keystone of the maintenance system is the user of equipment

kick (something) around	토의하다 (discuss; pass from one to another) We've got days to kick the plan around before deciding on a solution.
kickoff	개시 (beginning) The general went down to the th Infantry to watch the attack at the kickoff.
kick off	개시하다 (begin) The red flare was the signal to kick off the attack.
knock off or out	해치우다, 무력화하다 (kill; take out of action) We knock off the enemy artillery the first day.
know your way around	익숙해지다 (become familiar) with In each new job, it takes time to know your way around.
last-ditch	최후의 저항 또는 노력 (final resistance or effort) When we blew up all bridges in the rear, the enemy had to make a last-ditch effort.
last inning'(s), in the	최종 단계에서 (in the final phase) We launched our exploiting force in the lest innings of the campaign
last resort, as a	마지막 수단으로 (as a final means) The service elements of the division are used in combat only as a last resort.
last shift	마지막 교대조 (late tour of duty) I'm on duty form to ; that's the last shift.
layout	배치 (physical arrangement) The layout of the communications center must be efficient and convenient.
lay the cards on the table	상세하게 설명하다 (explain all the details) Before he commits the reserve company, he will have to lay the cards on the hable to the old man.
lay your hands on	확보하다 (secure; obtain) You've got to lay your hands on more ammunition somewhere.

leaning forward in their foxholes	공격할 태세다 (troops are ready and eager to begin the attack) At H hours all troops were leaning for ward in their foxholes.
leeway	여지, 여분, 여백 (margin; extra room or time) We had better start at once, for we have very little leeway for action
let off the hook	구제되다, 위기를 면하다 (to be let out of something; to be relieved) After a long explanation to has wife, John was let off the hook.
level, on the	공정한 (honest; trustful; fair; sincere) The new general has the reputation of always being on the level.
like falling off a log	쉬운 (easy) Passing that examination was like falling off a log.
live off(the fat of) the land	지역자원에 의지하다 (use local resources) Enemy policy is to live off the land as much as possible.
look down your throat	감제하다 (have dominant observation) From that hill mass, the enemy can look down your throat.
make a fine play	아주 뛰어나게 처리하다 (did something very well) The tank infantry team made a fine play through that hole.
make a stab at	시도하다, 해 보다(attempt; try; guess) I've never done that before, but I'll make a stab at it.
make good	합격하다, 성공하다 (be successful) We expect division units to make good on their combat readiness tests
makeshift	임기응변적인(improvised) This was a makeshift plan pending a complete analysis of the situation,

makeup	구성, 편성 (composition; arrangement) The makeup of that task force parallels that of a covering force.
manhour	인 시간의 노동량 (work of one man for hour) The meat of the matter is contained in the second paragraph.
meat of the matter	망치다 (ruin; spoil) You're in command now, so don't mess up your chances.
mess up	잘못 이해하다 (fail; misunderstand; completely mistake) He will miss the boat completely if he overestimates the enemy strategy.
miss the boat	소탕하다. 소진하다. (finish; complete) The seventh Infantry, following the 3rd Armored, will mop up the last enemy strategy.
mop up a notch	진급하다. (advance, as in promotion) Next month I'll move up a notch from G to G
move up	다수를 살상하다 (wound or kill promotion) Our interlocking fire will mow down the enemy infantry.
nail down	확정시키다, 딱 꼬집어 말하다 (to make final, firm, or) specific It was difficult to mail down the specific problem.
never say die	절대로 포기하지 마라 (never give up hope) We still have a chance, you know, so never say die.
nip and tuck	백중지세의 (very close; almost evenly matched) The fighting on our first objective was nip and tuck.
nose, on the	정확하게 (squarely; exactly; correctly) He answered that question on the nose.
no sweat	(쉬운 일이니깐) 걱정 마라 (don't worry; that which can be performed easily) No sweat, I can get this engine operating.
odds and ends	잡동사니 (remnants; small parts) The entire room was filled with odds and ends.

off balance, catch	미처 대비하지 못한 적을 덮치다 (find unprepared) By attacking at, we should be able to catch the enemy off balance.
off base	틀리다, 부정확하다 (wrong; incorrect) You are way off base on your time and space calculations.
offhand	대략 짐작컨대 (at first glance) Offhand, I'd say we'll capture prisoners on D day.
off the track	(시선 또는 주의를) 돌리다 (away; from; unaware) This maneuver will keep the enemy off the track of you main effort.
old man	상관; 중대장 (your senior officer; company commander) The old man doesn't always agree with his G-2, does he?
old standby	신뢰할만한 측근 (someone or something that has proved reliable) The commanding general is permitted to select his old standby as his chief of staff.
old stuff, that's	다 아는 것 (familiar to everyone) that's old stuff; it was in use at the outbreak of World War Ⅱ.
on the blink	제대로 작동하지 않다; 고장이다 (not operating properly) Why is this Universal Tester on the blink?
on the carpet (call)	꾸짖다 (to reprimand) the old man is going to put him on the carpet because of the serious mistake he made.
on the face of it	언뜻 보기에 (at first look) Your landing plan, on the face of it, looks OK to me.
on the fence, to sit	결정을 내리지 못하다. 망설이다 (to be undecided) Right now, he's sitting on the fence about the two courses of action.
on the hook	궁지에 (in a difficult position) You've got me on the hook; I just don't know the correct solution.

on the ropes	기진맥진한 (almost defeated) The enemy is on the ropes mow; keep hammering at him.
on the spot	결단을 요하는 (in a position requiring a decision) Your question puts mo on the spot, but here's my recommendation.
open-and-shut	확실한 (sure; certain; positive) They had an open-and-shut case against the accused.
out of hand	난장판인, 통제 불가능한 (out of control) During a bottleneck, traffic is nearly always out of hand.
out of the blue	아무 경고 없이 (without warning; unexpected) He dropped in at the house out of the blue.
our of the question	불가능한, 고려할 가치가 없는 (impossible; mot considered) That solution is completely out of the question.
out of the woods	위기를 벗어나서 (successful conclusion is in sight) In another day, we should be out of the woods.
out to get (someone)	본때를 보여주려고 벼르다. (to wait for the chance to do someone harm) Jack is really out to get Fred.
overall picture	전체적인 상황 (complete situation) I can brief you on the overall picture in minutes.
over his head, go	직속상관을 건너뛰어 호소하다. (appeal to higher authority) If he reuses your request, you ought to go over his head.
over my head	난해한 (difficult to understand; too complex) His explanation is over me head.
pain in the neck	짜증나는 사람(a boring person; a disturbing person) Mr. Jones is a pain in the neck.
paint a rosy picture	낙관적으로 묘사하다 (describe as ideal) He is not always reliable; he tends to paint a rosy picture too often.

pass the buck	책임을 전가하다. (to avoid responsibility by giving it to another) Don't try to pass the buck to me; that problem is yours.
pass the word down the line	전달하다 (communicate; convey the message) Pass the word down the line about the possible use of gas masks.
pay dearly	상대한 대가를 치르다 (costly) If we try to infiltrate, we'll pay dearly for it in lives.
payoff, the	결과, 결정 (the result; climax) The payoff of proper fire support coordination is the enemy's defeat.
pay off	좋은 결과를 낳다 (produce good results) Good maintenance ought to pay off in a reduced number of deadlined vehicles.
pep talk, a	사기를 높이는 말, 위로 (a talk to raise morale) What's the matter? you need p pep talk.
pet subject	가장 좋아하는 주제 (favorite topic) Military history is the colonel's pet subject.
pick and shovel work	어려운 과제, 업무 (difficult assignment(s)) The "Indians" on a staff of most of the pick and shovel work.
pick one out of thin air	별 근거 없이 추측하다, 고르다 (to guess or select) There are several solutions, so I'll pick one out of thin air.
piece of cake	쉬운 것 (any task or duty that is easy to perform) That examination was a piece of cake.
pin down the enemy	적을 고착시키다 (keep the enemy in on place) Our artillery will pin down the enemy indefinitely.
pinpoint	정확하게 겨누다, 위치를 포착하다 (to accurately locate information) You must pinpoint your targets when requesting artillery support.

plow though	빠르게 통과하다 (pass through quickly) An armored division cannot plow through heavily fortified positions.
plush job	편한 일, 사치 (easy job; a luxury) The division commanding general's trailer is a plush job for a home in th field,
point, there is no	아무 소용이 없다(there is no reason or use) There is no point in asking for it; there are none in stock
polish off	소탕하다 (defeat, complete) The mission of this unit is to polish off remaining resistance.
poor; bad; lousy; bum deal	좋지 않은 일 (happen) Going to school on Saturdays is a lousy deal.
pop up	일어나다 (happen) In a meeting engagement, anything can pop up, so be on the alert.
pound to pieces	격멸하다 (strike of attack heavily) Our artillery fire will pound to pieces any defense they establish.
premium, at a	진정 필요한(badly needed; difficult to obtain) Sleep is at a premium after a siege of hours.
pressed for time	시간에 쫓기다 (have very little time) A piecemeal attack is made when you're pressed for time.
pros and cons	장점과 단점 (advantages and disadvantages) Let's discuss the pros and cons of this course of action
pull off an operation	작전을 수행하다 (execute a maneuver) We can't pull off this operation without control of the sea approaches.
punch through	돌파하다 (penetrate) If we can punch though to Hill, We've got it made.

push off	개시하다 (begin) The predawn attack is scheduled to push off at 03:05 hours.
push on	진격하다 (advance) In the pursuit, mobile forces are to push on to cut off the enemy retreats
put a handle on	명확하게 말하다 (give something a name; be specific) Can you put a handle on that new gadget? I never heard of it.
put him on the spot	궁지에 처하게 하다 (place him in an awkward or a difficult position) A feint will put him on the spot because he won't know where out main effort is.
put your finger on it	명확하게 설명하다 (be specific; explain) Where's the gimmick in this? Can you put your finger on it?
rat race	악순환 (Seemingly senseless situation; hectic) That job at headquarters was a real rat race.
recap	요약하다 (Recapitulate; summarize) To recap briefly, our advance during the past hours has encountered stiff resistance.
recon	정찰, 수색 (Reconnaissance) This is a typical recon and fighter airfield of the tactical air force.
resume	요약 (A summary) A brief resume is the same as a recapitulation.
RHIP (rank has its privileges)	계급이 우선. It doesn't seem fair that the colonels get the best parking spaces, but RHIT.
right off the bat	즉시 (At once; initially) I've had a terrible morning. Our instructor gave us a quiz right off the bat.
ripe, the time is	적당한 시간(The correct time to act) The time is ripe for our counterattack, general.

road, hit the	떠나다 (To travel; to move on again) As soon as the graduation ceremony is over, I'm going to hit the road.
roadmap	지도 (Circuit or schematic diagram) Let's review the roadmap on this transmitter.
rough going, have	곤란하다, 어렵다 (Have difficulty) Our patrols on the mountains have rough going when communications fall.
rough time, a	어려운 시기, 곤란한 시기 (Difficult time) With his two assistants absent, the chief was having a rough time
round figures, in	대략 (Approximately) Our losses to date in round figures amount to about 105.
RTP (Read the problem or paper)	문제 또는 서류를 읽으면 The examination is easy if the students RTP.
run	지휘하다 (Command) The executive officer will have to run the company during the company commander's leave.
runaround	우유부단한 (Indecisive, evasive, or coy treatment) Although I kept asking for a direct answer, Joe kept giving me the runaround
run, on the	(적 따위가) 도주하여, 퇴각하여; 급히 (Going or moving quickly) We've got them on the run. Bring up the ammunition on the run, Joe
rundown, give a	요약설명을 하다 (Summarize; describe; brief) The G-2 can give a rundown at any time of all units organic and attached to the division.
run out of	고갈되다 (Come th the end of the available supply) After hours of driving, we ran out of gas.
run the show	책임지다, 담당하다 (Command; be in charge) Your job is to run the show at your assigned position.

saddled with
책임을 맡다 (Having the responsibility)
The military government officer is saddled with the problems of the refugee.

sack out
잠자리에 들다(To sleep; to go to bed)
I'm going to sack out after dimmer, I've had a hard day

scapegoat
(남을 대신한) 희생자 (One who is blamed)
The military police is usually the scapegoat if the traffic becomes congested.

scrap
싸움, 전투 (A fight)
Out main forces were engaged in a hot scrap all before noon.

scrape together
어렵게 긁어모으다. (Assemble with difficulty)
It required 4 days for our S-4 to scrape together enough rations for the forced march.

screwed-up
엉망인 (Spoiled; ruined; botched; confused)
That was a really screwed-up test.

scrimmaging
경미한 전투 (Light action)
There was only scrimmaging all along the main line of resistance.

see the light
이해하다 (Understand)
Thank you for your explanation, Now I begin to see the light.

set rule
정해진 규칙 (Answer; solution; regulation)
There's no set rule for each situation; use your imagination and initiative.

set up
준비하다 (Prepare; establish)
At the evening briefing, we are going to set up plans for the next day's mission.

setup
배치 (Arrangement; something very easy or certain)
The setup of the GG then must mot be dominated by the G section. the salient will prove to be a setup for our flanking fires.

shake off	털어 버리다 (Get rid of; lose) After an atomic attack, you have to shake off fear and resume the attack.
shakeup	개편, 변경 (Change) Because of the last defeat, there will be a shakeup in command.
shaky	자신 없는 (Weak or doubt) I'm a little shaky on my figures, kan; can you check them?
sharp	머리회전이 빠른(Intelligent; smart) A good soldier is always sharp.
shift, midnight	야간(Tour of duty from to) Activities are usually quiet during the midnight shift.
shoot!	주저 말고 하다 (Go ahead) If you have any questions, shoot!
shoot the works	전력을 다하다 (Use everything) When you commit your reserve, you shoot the works.
short cut	지름길(A quick way or method to save time) After experience in any job, you get to know the short cuts.
shorthanded	(사람이) 부한 (Lacking men) Our recent losses have left us shorthanded.
Short	고갈된, 다 써버린 (Ruined; worn out; used to) I'm short of ammunition, The patrol stopped short of the outpost about yards.
shot, take a	시도하다, 해보다 (Attempt; try) It's your turn to take a shock at it, Jim
shot in the dark	추측하다 (Guess) The old man doesn't take a shot in the dark; he acts on positive information.
shot up	피해를 입은 (Wounded; disorganized) My battalion was badly shot up in this attack.

showdown	정면대결 (Confrontation) This afternoon there will be a showdown between the various staff agencies.
showdown inspection	정밀점검 (Detailed inspection) The old man is going to have a showdown inspection tonight.
sit tight	대기하다 (wait; do not move) Just sit tight until his next move.
six by	6 륜차 (truck, six wheels driving) A six by has plenty of traction.
sixty four dollar question	가장 중요한 질문 (the most important question) Where are his reserves? That's the 64$ question.
size up the situation	상황을 파악하다 (make an estimate of the situation) The colonel quickly sized up the situation and then issued his orders.
sketchy	대략적인 (lacking in detail; vague; not clear) I couldn't act on such sketchy information
skin a cat	일을 끝내다. (accomplish mission, purpose, or job) Don't worry, I can fix this engine, for there is more than one way to
skip over	건너뛰다 (avoid mention of; omit) Allow me to skip over the next three items; now, how about evacuation?
siam into	맞부딪치다 (meet; engage violently) We must slam into the position with everything we have.
slang	속어 (expressions that are commonly used) If you don't savvy our slang, you can't participate in classroom discussions,
slide over	대충 넘어가다 (explain briefly) Don't slide over those matters; explain them at length.
slip up	잘못, 오류(make a mistake; error) When you are tired and nervous, you are likely to slip up on your job.

slug, a
총알 (abllet)
Slugs were flying thick and fast during the infiltration.

slugging match
격렬한 싸움 (an intense fight)
Capturing the final objective was a real slugging match.

smack into, run
정면으로 충돌하다(meet face to face)
A patrol may run smack into enemy ambushes.

snafu
약자표시대로 비정상 상태 (Situation Normal All Fouled Up)
Just as you said, the situation was really SNAFU.

snap decision, make a
급히 결정함 (act hastily; decide quickly)
Don't make a snap decision; take some time to study the situation.

sneak around the end
포위하다 (make an envelopment)
One course of action is to sneak around the end tonight.

snowball
점점 많아지다, 점점커지다 (become bigger)
During my absence, my work words to snowball.

snow job
달큼한 거짓말 (a pleasing but exaggerated speech)
It is amusing to listen to his snow job.

soft spot
약점 (weakness)
First fine the enemy's soft spot, then reinforce your initial success.

sort of
다소, 말하자면 (somewhat)
Every new job is sort of strange at first.

soup up
활기를 띠게 하다, 템포를 빠르게 하다 (increase)
We have to soup up the attack with all available artillery.

SOP (ESSOPEE)
예규, 현용운영절차
(Standing Operating Procedure; written directive describing correct procedures)
If you run the communications center according to the SOP, nothing will go wrong.

spec
기억하다 (memorize)
You can't spec everything; that's why we have an FM.

spell out	자세하게 설명하다 (to be specific; to explain step by step) Please spell out just what you mean so that I can understand.
spit it out; speak up	사실대로 말하다(to speak or say what you really thind) If you don't agree with me, speak up.
spot	발견하다 (discover; see; observe) Visual air reconnaissance is used to spot enemy movements.
squawk; beef; growl; bitch; gripe	불만, 불평하다 (a complaint; to complain) If a soldier doesn't squawk, he is mot happy.
stale, go	무관심해지다, 무기력해지다.(become careless or indifferent) After a long time at the same job, a man tends to go stale.
stalemate	무승부, 교착상태 (an impasse; a fight with on victory for either side)
stall off	핑계대어 시간을 끌다 (delay a decision) Your weapon is you best standby in a difficult spot.
standby	의지할만한 것 (sid; help; assistance; readiness) Be sure that you have a standby crew.
stay loose; hang loose	느긋하게 있어라 (remain flexible) Don't get excited; stay loose.
stick around	멀리 떠나지 않고 기다리다 (remain in the vicinity) Tell my aide to stick around till I get back.
stick it out	견디다 (endure it) Can you stick it out till reinforcement arrives?
stick your neck out	목숨을 걸다(take a chance) Don't stick your neck out without thorough reconnaissance.
stock answer	기존답안 (easy or predetermined solution) There's no stock answer in such a contingency.
straighten out	올바르게 고치다 (take corrective action) Go down there and straighten out that outfit, captain.

stretch your imagination	상상력을 동원하다(visualize with exaggeration) Can't you stretch your imagination to realize that this is possible?
strike out	실패하다 (fail) I struck out on that examination.
stumped	어찌할 바를 모르다 (perplexed) That has me stumped; what's your recommendation?
Sunday punch	작전 강타 (most punishing effort) Commitment of the reserve will be your Sunday Punch.
sweat it out	버티다 (wait uncertainly; endure until the end) We'll just have to sweat it out until we know his atomic capability.
swing a deal	성공적으로 이루다 (arrange successfully) Can you swing a deal to swap weapons at once?
tackle	(문제, 일 따위에)달라붙다, 씨름하다 (study; work on ; engage) We'll tackle this now and have the plan ready.
tailor-made	딱 맞게 (according to specification; correct) This tank is tailor-made for the job at hand.
take a crack at	시도하다 (attempt; try) Let me take a crack at it with my outfit.
take it, can	견디다 (can endure difficult situations) I admire his courage; he can take it.
take it easy	여유를 갖다 (relax) Don't let your problems upset you; instead, take it easy.
takeoff	이륙 (departure) Jet planes need a long runway, because their takeoff isn't fast.
take off	시작하다 (begin; stary; depart) It is approaching zero hour; take off now.
take over	인수하다 (take the place of; replace in command_ The colonel is to take over the regiment today; he'll run the outfit well.

take your pick	고르다, 선택하다 (choose) Here are two photos; take your pick and send me the other.
talk your of turn; shoot off	생각 없이 말을 하다 (blab; talk without discretion) He is famous for shooting off his mouth.
tangle with	교전하다, 맞부딪쳐 싸우다 (engage; come into contact with; fight with) A patrol is likely to tangle with several infiltrating groups.
tank dozer	앞에 날이 달려있는 탱크 (tank with a scraper blade) Do you know if there is an auxiliary engine in the tank dozer?
that's the rub	그것이 문제이다 (that's the difficulty) We can't get hold of any rations anymore; that's the rub
that's the way the ball bounces	운명이다 (that's fate) Our football team lost again; that's way the ball bounces.
that's your baby	네 책임이다. (something that is someone else's responsibility) Don't give me this project; that's your baby.
that will do	그만 하라 (that is enough; stop That will do, young man)
throw in the towel	항복하다 (to concede defeat; to surrender) The enemy suffered many casualties and were ready to throw in the towel.
throw out the window	제거하다, 삭제하다(eliminate; discard) This new information means we have to throw all earlier plans out the window.
thumbnail sketch	개요 (very general and brief descriptipn) G, I have only minutes for you to give me a thumbnail sketch.
tie down	구속하다, 제한하다 (limit; restrict; be explicit; finish preparation) The interrogation team tries to tie prisoners down on specific information that the G-2 wants.

tied up (다른 일로) 바쁘다 (Busy; detailed)
You can't see the old man; he's tied up

tie in with 일치하다 (agree with; support)
This information ought to tie in with what you found out.

time is of the essence 시간이 절박하다 (time is very importan)
Hurry that serial over the river; time is of the essence.

time-saver 시간을 절약하는 수단 (quick method to save time)
During briefing sessions, charts, graphs, and experience tables are great time-savers.

tip your hand 미리 정보를 주다 (give advance information)
Don't tip your hand by moving your reserves too early.

toehold, get a 침투하다 (penetrate)
Once you are able to get a toehold, set up the command post.

too much 이해하기에 너무 어려운
(beyond comprehension in either a good or bad sense)
That management course is too mush.

top man 우두머리 (the boss; the best man; the commander)
Ask LTC Brown; he's top man in this battalion.

touch on 언급하다, 간단히 말하다 (mention)
Now let's touch on supply; G-4 will be responsible for all rations.

tough time, have a 어려움을 겪다. (find it difficult to do)
The engineer company is likely to have a tough time putting in the bridge.

traffic jam 심한 교통체증 (a traffic difficulty)
Maximum flow and minimum necessary controls in a traffic plan will avoid traffic jams.

troop-lead the class 단계적으로 하나하나씩 설명해나가는 것
(instructor leads the class point by point and shows how to solve)
As a timesaver, the instructor may troop-lead the class though difficult points.

turn into	~이 되다 (become) Our attack could very easily turn into a route of the enemy.
turnover	인사변동 (number of people or items replaced during a specific time) There has been a large turnover of instructors.
two up and one back	2개 부대는 전방에, 2개 부대는 후방에 (two regiments in the front lines, one regiment in division reserve) In a normal defense position, we employ our infantry regiments tow up and one back.
umptyump	이러이런한 (something without a specific name or number) Where's he from? From the umptyunmp battalion, which just joined us.
Uncle Sam	미국, 미국연방정부 (The United States [from bewhiskered symbol of the United States]) Uncle Sam provides housing for many of the US students.
understudy	후임자 (one who trains for another person's position) I have been reassigned; LT Smith is my understudy.
under the gun	적의 관측 하에(dominated by enemy observation) Engineers putting up a bridge before th line is taken are working under the gun.
up [one's] alley	능력에 맞는, 전문에 속하는 (to be entirely within one's capabilities) I enjoy tactics; therefore, this course is definitely up my alley.
upset toe applecart	계획을 망치다 (ruin the plans) That undetected turning movement upset the applecart.
up-to date	최신정보, 최신상황 (the latest or newest information; current) Bring me up-to-the-date on what has happened in the last hours.

up-to the minute	최신정보, 최신상황 (up to the immediate present; the very latest) what's the up-to-the-minute news? No change? OK!
use your head	현명하게 생각하라 (Think intelligently) If a traffic snarl happens, sergeant, use your head and send the others on that bypass.
Whatyoumaycallit (watchamacallit)	정확하게 모를 때 사물은 부르는 말 (anything whose correct name you don't know) (watchamacallit) Give me that watchamacallit, the one by your left hand.
what's his face	정확하게 이름을 모를 때 사름을 말하는 호칭 (anything whose correct name you don't know) The course was taught by what's his face; you know who I mean?
watch your mouth	말조심을 하라 (be careful of what you say) When you talk to him, you must watch your mouth.
water under the bridge, it's	이미 지나간 사건 (a past event) You can't change that mistake, John; it's water under the bridge.
weasel out	교활하게 피하다 (to renege on a promise, usually for selfish reasons) Smith tried to weasel out of paying his debt.
what the hell?	도대체 무엇이냐? (an interjection indication anger, surprise, disappointment, etc)
what are you driving at?	무엇을 말하려고 하는 것이냐? (what do you mean?) What are you driving at by suggesting a change in leave quotas?
what gives; what's new; what's up?	무슨 일이냐? (a common greeting usually requiring no answer) Smith came up to our group and said, "what gives?"
whip	완벽하게 이기다 (beat; defeat) We must try to whip the enemy before winter sets in.

whipped	힘이 빠졌다, 졌다 (tried; defeated) I'm whipped; a fellow can't work hours without relief.
wind up an operation	끝맺다 (finish) This pursuit to the Rhine will wind up this operation.
work it out	해결하다 (solve) Here is the problem; try to work it out.
wrapped up in	몰두하다 (to devote full time or attention to) Robert is all wrapped up in his work.
Yankee	미국 북부지방 사람(an American from Northern States) He is a Yankee form new York.
you bet	당연하다 (surly; certainly; an expression of affirmation) Dave, would you like some dessert? You bet!
you've got something there	중요한 점이다 (you have an important there) I'd forgotten that point, but you've got something there.
zero hour	예정 행동개시 기각 (critical time) The zero hour for this attack will be 02:00 hours.

Abbreviations and Terms

약어 및 원어

A

AA: Ammunition Allocation 탄약할당
**assembly area 집결지
AAA: Anti-aircraft Artillery
대공포, 고사포, 대공포병
**AACG: arrival airfield control group
도착 비행장 통제단
AADC: Area Air Defense Commander
AAE: Army Aviation Element 육군 항공대
AAR: Air to Air Refueling 공중급유
**after-action review 사후검토
AASLT: Air Assault 공중강습
AAW: Anti-Air Warfare 대공전
AB: Air Base 공군기지
ABC: Atomic, Biological and Chemical
화생방
**ABCCC: airborne battlefield command and control center
공중전장 지휘통제본부
ABFDS: Aerial Bulk Fuel Delivery System
항공유류 분배장비
A/C: Aircraft 항공기
*AC: active component 현역 구성군

**ACA: airspace coordination area
공역협조지역

ACBLT: Air Contingency Battalion Landing Team
공수대대 비상착륙팀

[NOTE] * FM 100-5
** FM 7-20

ACC: Air Component Command
공군 구성군 사령부
Air Control Center 항공 통제 본부
ACCR: Air Component Command Regulation
공군구성군사 규정
**ACE: armored combat earthmover
전투공병 장갑차량
**ACL: allowable cargo load 인가탑재중량
ACofS: Assistant Chief of Staff 참모부장
ACP: Allied communications Publication
연합통신 간행물
**air control point 항공통제소
ACR: Armored Cavalry Regiment 기갑수색연대
ACSP: Assistant Chief of Staff, Personnel 인사참모부(장)
AD: Air, Defense 방공
ADA: Air Defense Artillery 방공포병

**ADAM: area denial artillery munition (antipersonnel)
야포발사식 지뢰 (대인)
**ADC: area damage control 지역피해통제
**ADM: atomic demolition munition 원자 폭발물
**admin/log: administrative/logistics
행정군수
**ADO: air defense office 방공장교
ADVON: Advance Party 선발대
AECC: Aeromedical Evacuation Control Center 항공의무 후송통제소
AEW: Airborne Early Warning 공중조기경보
AF: Air Force 공군
AFAS: Advanced Field Artillery System
차세대 자주포
AFATDS: Advanced Field Artillery Tactical Data System 전술적 사격지휘 자동화 장비
AFB: Air Force Base 공군기지
AFGWC: Air Force Global Weather Central (미)공군 중앙기상대
AFKN: American Forces Korea Network 주한미군 방송망
AFKTTL: Armed Forces Korean Tactical Target List
군 표적목록(주한미군)
AFOC: Air Force Operations Command
공군 작전사령부
AFSATCOM: Air Force Satellite Communications (미)공군 위성통신망
*AFSOF: Air Force special operations forces 공군 특수작전 부대
AFSPACECOM: Air Force Space Command (미)공군 우주사령부
**AG: Adjutant General 부관참모 (감)
AGI: Auxiliary General Intelligence 일반정보 보조
**AGL: Air Ground Operations 공지작건
AGS: Auxiliary Gathering Ship 정보 수집함
AGOSOP: Air Ground Operations Stand Operating Procedure 공지작전 예규
**AH: attack helicopter 공격용 헬기
AI: Air Interdiction 항공후방차단
AIC: Army Intelligence Command
육군 정보사령부
AIG: Address Indicator Group
수신처 표시군
AIS: Army Intelligence School
육군정보학교
**aj: antijamming 대전파 방해
AKTCAE: USA Korea Technical Control and analysis Element
주한 미 육군 기술통제분석반
ALB: Air Land Battle 공지전투
ALCOM: Alaska Command 알래스카사령부
**ALO: air liaison office 공군 연락장교
ALOC: Aerial Line of Communication
항공병참선
**alt: alternate 예비, 교대
ALT CP: CFC Alternate Command Post
연합사 예비지휘소
AM: Amplitude Modulation 진폭변조
AMC: Air Mobility Command
미 공수사령부
**(United States) Army Materiel Command
(미) 육군 물자사령부
AMEMB: American Embassy 미 대사관
AMHS: Automated Message Handling Equipment
자동 전문처리장비
AMO: Air Mobile Operation, 공중기동작전
*AMOPES: Army Mobilization and Operations Planning and Execution System
육군 동원 / 작전계획 및 실행체제

AMW: Amphibious Warfare 상륙전
ANGLICO: Air Naval Surface Fire Liaison Company 항공 함포연락중대
***AO**: Area of Operations 작전지역
AOA: Amphibious Objective Area 상륙 목표지역
***AOR**: Area of Responsibility 책임지역
****AP**: armored-piercing 철갑
APC: Armored Personnel Carrier 장갑인원수송차
****APDS**: armor-piercing discarding sabot 철갑축사
****APFSDS**: armor-piercing fin-stabilized discarding sabot 날개 안정 분리 철갑탄
APIR: Aerial Photograph Interpretation Report 항공사진 판독보고
APO: Army Post Office 군사우체국
APOD: Aerial Post of Debarkation 양륙 공항
ST: Air Refueling 공중재급유
***ARCENT**: Army Component to Central Command 육군 중부 사령부 구성군
***ARFOR**: Army forces headquarters 육군 본부
ARG: Amphibious Ready Group 상륙준비단
***ARSOF**: Army special operations forces 육군특수작전 부대
****ARTEP**: Air Raid Warming 공습경보
As: Anti-spoof 대 기만(對欺瞞)
ASAP: As Soon As Possible 가능한 빨리
ASARS: Advanced Synthetic Aperture Radar System 개량 영상정보 레이더 체계
ASAS: All Source Analysis System 종합정보 분석체계
ASAT: Anti-satellite 대 위성
ASC: AUTODIN Switching Center 자동 전신타자 교환대
ASCG: Ammunition Supply Coordination Group 탄약지원협조단
ASG: Area Support Group 지역지원단
ASIC: All Source Intelligence Center 전출처 정보센터
ASOC: Air Support Operations Center 항공지원작전본부
ASP: Ammunition supply Point 탄약보급소
ASR: Ammunition supply Route 탄약보급로
ASUW: Anti-surface Warfare 대 수상전(對水上戰)
Anti-Submarine Warfare 대 잠전(對潛戰)
****at**: antitank 대전차
ATACMS: Army Tactical Missile System 지대지 유도무기
ATC: Army Transportation Command 육군 수송사령부
ATG: Amphibious Task Group 상륙기동단
****ATGM**: antitank guided missile 대전차 유도탄
ATM: Asynchronous Transmission Mode 비동기식 전달모드
ATSP: Air Terminal Supply Point 항공단말보급대
****ATP**: ammunition transfer point 탄약전환(인도)지점
ATT: Army Training Test 육군훈련시험
****auth**: authorized 인가된
AUTODIN: Automatic Digital Network 전문 데이터자동 송수신망(범세계 통신망)
AUTOVON: Automatic Voice Network 자동 정화망
****AVLB**: armored vehicle-launched bridge 장갑차량 추진교량
Avn: Aviation 항공

AVNDDE: Aviation Brigade 항공여단
AVRE: Assault Vehicle Royal Engineers 습지극복 매트
***AWACS**: Airbrne Warning and Control System
공중경보 및 통제체제 (공중조기경보기)
AWF: All-Weather Fighter 전천후 전투기
AWN: Automated Weather Network 자동 기상망
AWS: Air Weather Service 항공기상근무(대)
****as**: azimuth 방위각

B

BAI: Battlefield Air Interdiction
전장항공차단
BATTLE: Battalion Analyzer & Tactical Trainer for Local Engagements
대대전투분석 및 전술훈련용 모의 전투
BCE: Battle Coordination Element
전투 협조반
BCIS: Battlefield Combat Identification System 전장 전투 식별 체계
BCTP: Battle Command Training Program 전투지휘참모훈련
BCWG: Battle Coordination Working Group 전장협조실무단
***BDA**: Battle Damage Assessment
전투피해평가
**bomb damage assessment 폭격피해평가
***BDAR**: battle damage assessment and repair
전투 피해평가 및 수리
Bde: Brigade 여단
B/L: Basic Load 기본 휴대량
BMNT: Beginning Morning Nautical Twilight
해상박명초
Bn: Battalion 대대
BO: Black put 등화관제
BOQ: Bachelor Officers' Quarters
독신장교숙소
****BOS**: battlefield operation system
전장운영체제
****BSA**: brigade support area 여단지원지역
BSL: Bomb Safety Line 폭격안전선
****btry**: bettery 포대
BTCS: Battalion Tactical Computer System
포병대대 사격지휘 자동화 장비
BWP: Basic War Plan 기본전쟁계획
BY : Budget Year 회계 년도

C

***C2**: Command and Control 지휘 및 통제
C2: Combined Intelligence Staff
연합사 정보참모부
C3: Command, Control, Communication
지휘, 통제 및 통신
C3: Combined Operations Staff
연합사 작전첨모부
***C3CM**: C3 Countermeasures
지휘, 통제, 통신 및 대응책
***C3I**: Command, Control, Communications and Intelligence 지휘, 통제, 통신 및 정보
***C3IC**: coalition, coordination communications, and intergration center
C4: Combined Logistics staff 연합사 군수참모부
C4I: Command, control, communications, computers and Intelligence 지휘, 통제, 통신, 컴퓨터 및 정보
C5: Combined Policy and Plans Staff
연합사 기획참모부
C6: Combined communications Staff

연합사 통신전자참모부
***CA**: Civil Affairs 민사
Convening Authority 소집권한
**Counter Air 대공(對空)
CAC: Comb Arms Command
전투병과교육사령부
CAC: Combat Air Command
(북한)항공전단사령부
CACC: Commander, Air Component Command
공군구성군사령관
CAF: Combined Aviation Force
연합육군항공부대
CAFSOB: Combined Air Force Special Operations Base 연합공군 특전기지
CADSOCC: Combined Air Force Special Operations Coordination Center 연합공군 특전협조 본부
CALO: Combined Airlift Office
연합공수사무소
CAN-K: Combat Assessment Model-Korea
전투평가 모델
CAP CORPS: Capital Corps 수도군단
****CARP**: computed air release point
공중 폭탄투하 지점
CAS: Close Air Support 근접항공지원
CASIC: Combined All Source Intelligence Center
연합 전출처 정보센터
CASOP: Crisis Action System Operating Procedures 위기조치예규
CAT: Crisis Action Team 위기조치반
CATF: Commander Amphibious Task Force
상륙기동부대사령관
****cav**: cavalry 기갑부대

CBNL Chemical, Biological and Nuclear 화학, 생물, 핵
CBR: Chemical, Biological and Radiological 화생방
CBSSOP: Combined Battle Staff Standing Operating Procedures
연합전투참모단 예규
****cbt**: combat 전투
****CBU**: cluster bomb unit 인명 사살용 폭탄
CCL: Civilian Control Line
민간인 통제선(민통선)
CCPDSR: Command Center Processing and Display System Replacement
지휘소 처리 및 시현장비 교체
CCSC: Combined Communications Support Center 연합통신지원본부
CC Seoul: Command Center Seoul
서울지휘소
****CCT**: combat control team 전투 통제반
CCTV: Closing Circuit TV 폐쇄 회로 TV
C&G: Cover and deception Operations
엄폐 및 기만작전
C-Day: The day on which a deployment operation commences or is due to commence
증원군 전개 개시(예정)일
CDC: Capital Defense Command
수도방위사령부
Cdr: Commander 사령관, 지휘관
CE: Command Element 지휘반, 지휘부
Communications Electronics 통신전자
CEB: Combat Engineer Battalion
전투공병대대
CEMS: Construction Equipment Multipurpose Section 다목적 굴착기

CEN: Communication Engineer 통신공병
***CENTAG**: Central Army Group (NATO) 집단군 (나토)
CEOI: Communications-Electronics Operation Instruction 통신전자 운용지시
****CEV**: combat engineer vehicle 전투공병차량
CEVG: Combat Evaluation Group 전투평가단
CFACC: Combined Force Air Component Commander 연합군 공군구성군사령관
CFC: Combined Forces Command 연합사
CFCA: Assistant Chief of Staff, C-1 연합사 인사참모부 (연인)
CFCB: Assistant Chief of Staff, C-2 연합사 정부참모부 (연정)
CFCD: Assistant Chief of Staff, C-3 연합사 작전참모부 (연작)
CFCE: Assistant Chief of Staff, C-4 연합사 군수참모부 (연군)
CFCF: Assistant Chief of Staff, C-5 연합사 기획참모부 (연기)
CFCH: Assistant Chief of Staff, C-6 연합사 통신전자참모부 (연통)
CFEN: Assistant Chief of Staff, Engineer 연합사 공병참모부 (연공)
CFHC: HQ Commandant 연합사 본부사령실 (연본)
CFJA: Judge Advocate 연합사 법무실 (연법)
****CFL**: coordinated fire line 사격 협조선
CFPA: Public Affairs Office 연합사 공보실 (연공보)
CFSC: Secretary Combined Staff 연합사 비서실 (연비)
CGCC: Commander, Ground Component Command 지상군구성군사령관
****chem**: chemical 화학
CHOP: Change of Operational Control 작전통제권 이양
chosen Soren: General Federation of Koreans Residing in Japan 조총련
C/HUMINT: Counter-Human Intelligence 대 인간정보
CI: Civilian Internees 민간인 억류자
Command Information 지휘첩보
Counter-Intelligence 대 정보
CIA: Central Intelligence Agency 중앙정보부
CIDS: Command Information Display System 지휘정보 전시체계(화상회의)
C/Imint: Counter-Imagery Intelligence 대 영상정보
***CINC**: Commander in Chief 사령관
CIOC: Combined Intelligence Operation Center 연합정보 운영센터
CIS: Commonwealth of Independent States 독립국가연가
***CJCS**: Chairman, Joint Chief of Staff (US) 합참의장
CJIB: Combined/Joint Information Bureau 연합합동 정보국
CJRFL: Combined Joint Restricted Frequency List 연합합동 제한주파수 목록
***CJTF**: commander, joint task force 합동 특수임무부대 사령관
CL: Combat Loading 전투 적재
Class I: Supplies - Subsistence 1종 - 식량
Class II: Supplies - Clothing and Individual Equipment 2종 - 피복 및 개인장구
Class III: Supplies - Petroleum Oils, and Lubricants

3종 - 육류

ClassⅣ: Supplies - Construction Material 4종 - 축성자재

ClassⅤ: Supplies - Ammunition 5종 - 탄약

ClassⅥ: Supplies - Personal Demand Items 6종 - 피엑스 품목

ClassⅦ: Supplies - Major End Items 7종 - 주요 완제품

ClassⅧ: Supplies - Medical Material 8동 - 의무물자

ClassⅨ: Supplies - Repair Parts 9종 - 수리부속품

CLC: combat Lohistics Center 전투군수본부

CLCA: Combined Logistics Coordinating Agency 연합군수협조기구

CLCC: Combined Logistics Coordinating Committee 연합군수협조위원회

CLF: Commander Landing Force 상륙군사령관

****CLGP**: cannon-launched guided projectile 레이저 유도 폭탄

CM: Court-Martial 군법회의

CMC: Cheyenne mountain Complex 북미 우주방공사령부가 위지한 기지

****cmd**: command 사령부

CMEC: Combined Marine Expeditionary Corps 연합해병원정군단

CMFC: Combined Marine Forces Command 연합해병사령부

CMIC: Combined Military Interrogation Center 군 합동심문소

CNLCC: Combined Marine Logistic Coordination Committee 연합해병 군수협조위원회

****CMO**: civil-military operations 민사작전

CMV: Combat Mobility Vehicle 전투공병 차량

CNCC: Commander, Naval Component Command 해군구성군사령관

CNSWTG: Combined Naval Special Warfare Task Group 연합해특단

****co**: company 중대

****CO**: commanding officer 지휘관

****COA**: coures of action 방책

COB: Co-located Operating Bases 공동운영기지

COBRA: Assault Helicopter AH-1 코브라헬기

***COCOM**: Combatant Command 전쟁지휘

COG: Component Operations Group (해군)성분전단

COIN: Counterinfiltration, Counterinsurgency 대간첩

COMCMFC: Commander, combined Marine Forces Command 연합해병사령관

COMCUWTF: Commander, combined Unconventional Warfare Task Force 연합특전사령관

CONINT: Communications Intelligence 통신정보

COMJAM: Communications Jamming 통신방해

****comm**: communication 통신

****CONNEX**: communications exercise 통신연습

COMMSUCCESS: Communications Success Report 통신개통보고

***COMMZ**: Communications Zone 병참

지대
COMSEC: Communications Security 통신보안
CONSPOT: Communications Spot Report 통신상황 보고
COMUSNAVFOR: Commander, Naval Forces 미 해군부대사령관
COMUWTF: Commander, Unconventional Warfare Task Force 특전사령관
CONOPS: Concept of Operations 작전개념
CONPLAN: Concept Plan 개념계획
***CONUS**: Continental United States 미 본토
COP: Command Observation Post 지휘관측소
CORCEN: Correlatin center 분석실 센터
**COSCOM: corps support command 군단 지원 사령부
CP: Check Point 검문소
**command post 지휘소
CPOTF: Combined Psychological Operations Task Force 연합 심리전특수임무대대
CPSC: Combined Petroleum Support Center 연합 유류지원본부
CPX: Command Post Exercise 지휘소연습
C-R: Confidential-Republic of Korea / United States
한-미 3급 비밀
***CRC**: Combat Reporting Centers 중앙관제소
CRDL: Critical Requirement Deficiency List
주요 소요부족 목록
C-RELROK: Confidential-Releasable Republic of Korea 한국공개가능 3급 비밀
CRP: Combat Reporting Posts 지방관제소
**CS: combat support 전투지원
CSAR: Combat Search and Rescue 전투탐색 및 구조
CSCT: Combat support Coordination Team 전투지원 협조반
CSE: Combat Support Equipment 전투지원장비
C/SFOB: Combined Special Forces Operations Base 연합특수작전기지
C/SIGINT: Counter-signals Intelligence 대 신호정보
CSLCK: COMSEC Support Logistics Center/Korea
주한 통신보안 군수지원본부
CSP: Communications Support Processor 통신지원절차
**CSR: Controlled Supply Rate 통제보급률
CSS: Central Security Service 중앙보안기관
**combat service support 전투근무지원
CTAPS: Contingency Tactical Aircontrol System Automated Plannings System
전술항공 통제 자동화 기획체계
CTAT: Combined Terrain Analysis Team 연합 지형분석반
CTB: Combined Targeting Board
종심전투 통합위원회
CTF: Combined Task Force
연합 기동부대
CTMC: Combined Transportation Movement Center 연합 수송이동본부
CULT: Common User Land Transportation 고로수송(지상)

CUWTF: Combined Unconventional Warfare Task Force 연합특전사
CVBG: Aircraft Carrier Battle Group 항공모함 전투단
CVN: Multi-Purpose Aircraft Carrier (Nuclear Propulsion) 다목적 핵추진항공모함
**CW: continuous wave 지속파
*CZ: combat zone 전투지대

D

*DA: Department of the Army 미 육군성
DAG: Division Artillery Group 사단 포병군
*DAO: defense attache office
DBSL: Deep Battle Synchronization Line 종심전투 통합선
DCA: Defensive Counter-air 방어제공
DCS: Defense Communication System 국방통신체계
DCS: Defense Courier Service (미)국방 전령근무
D-Day: The day on which an operation commence or is due to commence 작전개시(예정)일
DDN: Defense Data Network (미)국방부 데이터 통신망
DECCO: Defense Commercial Communications Office 국방 상용통신소
Decon: Decontamination 제독(除毒)
DEFCON: Defense Readiness Condition 방어준비태세
DF: Direction Finding 방향탐지
DFSC: Defense Fuels Supply Center (미)국방부 유류보급본부
DIA: Defense Intelligence Agency (미)국방부 정보국
DIEOB: Defense Intelligence Electronic Order of Battle (미)국방부 정보전자전 전투서열
DISA: Defense Information Systems Agency (미)국방 첩보체계국
**DISCOM: division support command 사단 지원사령부
**Div: Division 사단
DIVLEW: Division Level Wargame Model 사단 워게임 모델
DMA: Defense Mapping Agency (미)국방부 지도제작창
DMD: Digital Message Device 관측제원 입출력기
DMED: Digital Message Entry Device 고속 문서 처리기
DMSP: Defense Meteorological satellite Program (미)국방부 기상위성계획
DMZ: Demilitarized Zone 비무장지대
**DOD: Demilitarized Zone (미)국방부
**DODAC: Department of Defense ammunition code 국방부 탄약기호
DODIC: Department of Defense Item Code (미)국방부 식별부호
DOS: Day of Supply 보급일수
DP: Detainees 억류자
DPER: Daily Personnel Effectiveness Report 일일병력보고
**DPICM: dual-purpose improved conventional munition 이중목적 개량탄
**DS: Direct Support 직접지원
DSC: Defense security Command 국군보안사
DSCS: Defense Satellite Communications System (미)국방 위성통신체계
DSCSOC: Defense Satellite

Communications System (미)국방 위성통신체제 운용본부

DSN: Defense Switched Network (미)국방부 전화교환통신망

DSP: Defense Support Program 방어지원계획

Dtd: Dated(As in the date an item is published) (발행)일자

****DTG**: date-time group 일시군

DTSS: Digital Topographic Support System 수치 비형분석 지원체계

DTV: Driver's Thermal Viewer 조종수 열상 관측경

DZ: Drop Zone 낙하(투하)지대

E

EA: Electronic Attack 전자 공격

***EAC**: Echelons Above Corps 군단급 이상 제대

***EAD**: echelons above division 사단급 이상 제대

ECCN: Electronic Counter-Counter-Measures 전자방해 방어대책

****ECCM**: electronic counter-countermeasure 전자방해 방어대책

****ECM**: Electronic Counter-Measures 전자방해대책

E&E: Escape and Evasion 도피 및 탈출

EEI: Essential Elements of Information 해상박명종

EEPI: Essential Elements of Friendly Information 우군 첩보기본요소

****ELINT**: Electronic Intelligence 전자정보

****ELSEC**: electronic security 전자보안

EM: Enlisted Man 사병

EMCON: Emission Control 전파발사 통제

****EMP**: electromagnetic pulse 전자기 펄스

EO: Electronic Optical 전자 광학

EOB: Enemy Order of Battle 적 전투서열

EO-X: Electronic Optical - X 전자 광학 - X

EP: Electronic Protection 전자 방호

****EPW**: Enemy Prisoner of War 적포로

ER: Electronic Reconnaissance 전자 수색

ES: Electronic Warfare Support 전자전 지원

ESB: Engineer Support Battalion 공병지원대대

****ESM**: Electronic Warfare Support Measures 전자전 지원방책

****ETA**: estimated time of arrival 예상 도착시간

EUSA: Eighth United States Army 미8군

****EW**: Electronic Warfare 전자전

EWL: Engineer Workline 공병작업 한계선

EWO: EW Officer 전자전장교

****EZ**: extraction zone 퇴출지역, 탑재지역

F

F/A: Fighter/Attack 전투기/공격기

****FA**: Field Artillery 포병

***FAAD**: forward area air defense 전방지역 방공포대

****FAC**: Forward Air Controller 전방항공통제관

****FASCAM**: Family of Scatterable Mines 살포식 지뢰체제

****FAST**: forward area support team 전방지역 지원팀

Forward Area Support Terminal 전방지역 지원단말
**FAX: facsimile 전송사진, 복사전송장치
FCG: Full Combat Gear 완전군장
FCO: Fire Control Officer 화력통제장교
**FCL: fire coordination line 화력 협조선
**FD: fire direction 사격지휘
**FDC: Fire Direction Center 사격지휘소
FDO: Flexible Deterrent Option 신속억제방안
**FEBA: Forward Edge of the battle Area 전투지역전단
*FEMA: Federal Emergency Management Agency 연방 위기관리 기관
**FFA: free fire area (포병) 자유사격지역
*FID: foreign internal defense
5D: Demand Driven Direct Digital Dissemination 영상분배 네트워크
FIP: Force Improvement Program 전투력발전계획 (율곡계획)
**FIST: Fire Integrated Support Team 화력통합지원반
FIST-V: Fire Support Team-Vehicle 관측반용 장갑차
**fld: field 야전
FLIR: Forward Looking Infrared 전방감시용 적외선 장비
**FLOT: Forward Line of Own Troops 전선
FLT: Front Line of Trace 전방접촉선
FLTSATCOM: Fleet Satellite Communi cations 함대 위성통신
**FM: Frequency Modulation 주파수 변조방식
*field manual 야전교범
FMS: Foreign Military Sales 대외군사판매
*FMSP: Foreign Military Sales Program 외국 군사판매 계획
**FO: forward observer 관측장교
FOMOA: Follow-in Memorandum of Agreement 후속 합의각서
*FORSCOM: United States Forces Command 미군 사령부
**FPF: Final Protective Fire 최후저지사격
**FRAGO: fragmentary order 단편 명령
*FRAGPLANL: fragmentary plan 단편 계획
FROG: Free Rocket Over Ground 프로그 미사일 (지대지)
FROKA: first Republic of Korea Army 한국육군 제 1군
FRP: Fiberglass Reinforced Plastic 유리섬유로 강화된 플라스틱
**FS: fire support 화력지원
FSA: Fire Support Area 화력지원지역
**FSC: fire support coordination 화력지원협조
**FSCC: Fire Support Coordination Center 화력지원협조소
FS Cell: Fire Support Cell 통합 화력지원반
**FSCL: Fire Support Coordination Line 화력지원 협조선
**FSCOORD: fire support coordinator 화력지원 협조관
**FSE: Fire Support Element 화력지원반
**FSO: fire support officer 화력지원장교
*FSSG: First Service Support Croup 제1근무 지원단
ft: feet 피트
FTX: Field Training Exercise 야외훈련연습

G

GBU: Guided Bomb Units 항공유도 폭탄류
GCC: Ground Component Command 지상구성군
GCMCA: General Court-Martial Convening Authority 보통군법회의 소집기구
**GEMSS: Ground Emplaced Mine Scattering System 지상살포식지뢰
GEOREF: The World Geographic Reference System 범세계 지리참조 (좌표)
GHz: Gigahertz 기가헤르쯔 (전파단위)
**GL: Grenade Launcher 유탄발사기
**GLLD: ground laser locator designator 지상감시 레이더
GLSB: GeneralLogistics Support Base 일반 군수지원기지
GMIT: Ground Moving Intercept Team 지상이동 차단팀
*CMR: graduated mobilization response
GOP: General Outpost 일반초소
GP: Guard Post 감시초소
**general purpose 일반목적
GPS: Global Positioning System 범세계 위치식별체계
GR: Graves Registration 영현등록
**GS: General Support 일반지원
**GSA: General Services Admini stration 미 연방 조달청
GSM: Ground Station Modules 지상 단말처리소 Ground Station Mode 지상 수신 모드
**GSR: ground surveillance radar 지상감시 레이더
**GZ: ground zero 지상원점, 폭팔 중심지점

H

HAB: Heavy Assault Bridge 중(重)강습용 가교
HARTS: Hardened Artillery Site 갱도포 병진지
HAWK: Homing All the Way Killer 호크
H-Camera: Homeland Defense Reserver 고감도 카메라
HDRF: Homeland Defense Reserve Forces 향토예비군
**HE: high explosive 고폭탄
**HEAT: high-explosive antitank 대전차 고폭탄
HEAT_MP: High Explosive Antitank Multi Purpose 다목적 대전차 고폭탄
**HEDP: high-explosive dual-purpose 이중목적 고폭탄
HELR: High Explosive Infrared 고성능 적외선
**HEI-T: high-explosive incendiary-tracer 고폭 소이 예광탄
HERC: Nike-Hercules 나이키-허큘리스
HET: Heavy Equipment Transport 중장비 수송차량
**HF: High Frequency 고주파
**HHC: headquarters and headquarters company 본부 및 본부중대
H-Hour: The specific time at which an operation commences or is due to commence 작전개시시간
*HIMAD: High-Medium Air Defense 중, 고고도 방공무기체계
HIMARS: High-Mobility Artillery Rocket System 고기동 다련장 로켓
HIRES/2: Second Generation High Resolution 차세대 고감도형

HJ: Honest John 어네스트존 미사일
H/L: Hot Line 직통전화
HMCT: Highway Management Control Teams 육로이동관리반
HMI: Host Message Interface 주전문 연동
****HMMWV**: high-mobility, multipurpose wheeled vehicle 고기동성 다목적 차륜 차량
HNS: Host Nation Support (주둔국)지원
HNSA: Host Nation Support Agreement (주둔국)지원협정
****HQ**: headquarters 사령부, 본부
HQ COMDT: HQ Commandant 본부사령
HRDIV: Homeland Reserve Division 향토예비사단
****HSS**: health service support 위생근무 지원
HTACC: Hardened Tactical Air Control Center 전술항공 통제본부
HTCCF: Hardened Taegu Command and Control Facility 대구지휘통제시설
HQ: Headquarters 사령부
HR: Homeland Reserve 향토예비군
HRD: Homeland Reserve Division 향토사단
HRE: Han River Estuary 한강하구
HRFCC: Han River Flood Control Center 한강 홍수통제소
****HUMINT**: Human Intelligence 인간정보
Hvy: Heavy 중(重)

I

****IAEA**: International Atomic Energy Agency 국제원자력기구
****IAW**: In Accordance With …에 의거
IBM: International Business Machine IBM컴퓨터 장비
ICBM: Intercontinental Ballistic Missile 대륙간탄도탄
****ICM**: Improved Conventional Munition 개량탄
ICR: Intelligence Collection Requirements 정보수집요구
ICTT: Improve Command Tactical Terminal 전술지휘 단말
ID: Infantry Division 보병 사단
Identification card 식별 카드
IDH: Indian Head 인디안헤드(미2사단의 별칭)
IDX: Improved Dispersed Explosive 기화 폭약
****IED**: imitative electronic deception 모의 전자기만
****IEW**: intelligence and electronic warfare 정보 및 전자전
IFF: Identification Friend of Foe 피아 식별
IFV: Infantry Fighting Vehicle 보병 전투용 장갑차
****IG**: inspector general 감찰감, 감찰병과
IH: Improved Hawk 개량호크미사일
IIR: Imagery Intelligence Report 영상판독보고
IM: Intercepter Missile 요격유도탄
***IMETP**: International Military Education and Training 국제 군사교육 및 훈련계획
IMHS: Interactive Message Handling System 상호 전문처리체계
****IMINT**: Imagery Intelligence 영상정보
****in**: inch(es) 인치
****IN**: infantry 보병
Inf (or IN): Infantry 보병

INH: Improved Nike-Hercules 개량나이키허큘리스
**INS: inertial navigation system 관성항법장치
INTREP: Intelligence Report 정보보고
**INTSUM: Intelligence Summary 정보요약
IP: Incentive Pay 위험수당
**initial point 기점
IPAC: Intelligence Center Pacific 태평양 정보본부
**IPB: intelligence Preparation of the Battlefield 전장정보분석
IPDS: Inland Petroleum Discharge System 지상유류 분배장비
IPF: Integrated Processing Facility 통합처리시설
IPIR: Initial Photographic Interpretation Report 최초 사진판독 보고
**IPW: Interrogation Prisoner of War 포로심문
**IR: Information Requirements 첩보요구
Infrared 적외선
Intelligence Requirements 정보요구
IRBM: Intermediate Range Ballistic Missile 중거리탄도탄
**ISB: intermediate Staging base 중간 대기기자
ISSA: Inter-Service Support Agreements 각군 상호지원협정
*ITEP: intelligence and tactical exploitation processor
ITM: Intercepter Tactical Missile 전술요격 미사일
ITP: Integrated Tasking Order 통합 임무 명령서
ITW/AA: Integrated Tactical Warning and Attack Assessment 통합 전술경보 및 공격평가
I&W: Indications and Warnings 징후 및 경보

J

J-7: ROKMND/JCS Civil Affairs/PSYOP Directorate 한국 합참 민사심리전참모부
JA: Judge Advocate 법무
**JAAT: joint air attack team 합동 공중 공격팀
**JAG: judge advocate general 법무감
JCCC: Joint Communications Control Center 합동통신 통제본부
JCIS: Joint Command Information System 연합 지휘정보관리체계
*JCS: Joint Chiefs of Staff 합동참모본부
*JFC: joint force commander 합동부대 지휘관
J-Fire: Joint Application of Firepower 통합 화력운용
*JFLCC: joint force land component commander 합동부대 지상구성군 사령관
JFMO: Joint Frequency Management Office 합동 주파수 관리사무소
JICPAC: Joint Intelligence Center Pacific 미 태평양 합동정보본부
JMA: Joint Mobilization Augmentation 전시 증원표
JMFC: Republic of Korea/United States Joint Military Frequency Committee 한미 합동 군사주파수 관리위원회
JMP: Joint Manpower Program 합동인력

계획

JMRO-K: Joint Medical Regulation Office - Korea 주한미합동 의무조정실

*__JOA__: joint operations area 합동작전 지역

JOG-G: Joint Operations Planning and Execution System 합동기획 및 시행체제

JOPS: Joint Operation Planning System 합동작전 계획체제

JOTS: Joint Operational Tactical System 합동작전용 전술경보체제

JRCC: Joint Rescue Coordination Center 합동구조협조본부

JSA: Joint Security Area 공동 경비구역

JSCAP: Joint Strategic Capabilities Plan 합동 전략능력기획서

*__JSCP__: Joint Strategic Capabilities Plan 합동전략 능력기획

**__JSEAK__: joint suppression of enemy air defenses 합동 적방공 제압

JSOC: Joint Staff Operations Agency 합동참모작전본부

*__J-STARS__: Joint Surveillance and Target Attack Radar System 합동 감시표적 공격 레이더 체계

FTB: Joint Transportation Board 합동 수송위원회

*__JTF__: Joint Task Force 합동 특수임무부대

K

KAC: Korea Analysis Center 한국 분석센터

KAIS: Korean Air Intelligence System 한국 항공정보체제

KAL: Korean Air Lines 대한항공

KALCC: Korean Airlift Control Center 한국 공수통제본부

KATUSA: Korean Augmentation to United States Army 카투사

KB: Knowledgeability Brief 심문첩보 보고

KCA: Korea Contracting Agency (US) 주한 계약담당관

KCITS: Korean Compressed Imagery Transmission system 한국군 영상 전송 체계

KCOIC: Korean Combat Operations Intelligence Center 한국 전투작전 정보본부

KCRCC: Korea Combined Rescue Coordination Center 한국 연합구조협조본부

KDSS: Korean Combat Support System 한국 전투지원체제

KDIA: Korean Intelligence Agency 한국 정보본부

KEPCO: Korean Electric and Power Company 한국 전력공사

KFS: Korean Flag Ship 한국 국적 선박

KFU: Korean Forecast Unit 한국 기상부대

KHz: Kilohertz 킬로헤르쯔 (주파수 단위)

**__KIA__: Killed in action 전사(자)

KIDA: Korean Institute for Defense Analysis 한국국방연구원

K-IFS: Korea Intelligence Fusion System 한국정보융합체계

KMPA: Korean Maritime and Port Administration 한국 해운항만청

KNR: Korea National Railroad 한국 철도공사

KOSIDS: Korea Satellite Imagery Dissemination System 한국 위성화상

전송체제
KPA: Korean People's Army 조선인민군
K-SAM: Korea Surface To Air Missile 한국형 단거리 대고 유도탄
KT: Korea Telecom 한국통신(주식회사)
KTACS: Korean Tactical Air Control System 한국 전술항공통제체제
KTMF: Korean Tactical Map Facility 한국 육군지도창
KTMR: Korea Traffic Management Regulation 한국 교통관리규정
KTO: Korean Theater of Operations 한국 전구작전

L

LAA: Limited Access Authorizations 제한접근 인가(인원보안)
LAN: Local Area Network 근거리 통신망
****LAPES**: low-altitude parachute extraction system 저공추출 낙하산체제
****LAW**: light antitank weapon 경 대전차 무기
****lb**: pound(s) 파운드
****LBE**: load-bearing equipment 군장류
****LC**: line of contact 포착선
LCG: Logistics Coordinating Group 군수협조단
****LD**: line of departure 공격개시선
****ldr**: leader 지휘자
LIB: Light Infantry Brigade 경보병여단
LNO: Liaison Officer 연락장교
****LO**: liaison officer 연락장교
****LOC**: Line of Communication 병참선
LOCE: Linked Intelligence Operation Center Europe 센서로부터 첩보수신 근실시간대 정보융합 및 자동전파체제
****log**: logistics 군수
***LOGCAT**: Logistics Civil Augmentation Program
****LORAN**: long-range aid to navigation 장거리 항법장치
****LOS**: line of sight 조준선, 시선, 레이더
***LOTS**: logistics-over-the-shore 해안상륙 군수지원
LP: Listening Post 청음초
LP&P: Logistics Policies and Procedures 군수방침 및 절차
LRE: Laser Range Finder 레이저 거리 측정기
****Lt**: Light 가벼운, 경(經)
LT: Landing Team 상륙팀
LW: Light Warning 등화관제
****LZ**: Landing Zone 착륙(상륙)지점

M

****m**: meter(s) 미터
M: Mechanized 기계화
MAA: Military Armistice Agreement 군사정전협정
****MAC**: maintenance allocation chart 정비 할당표
MAJCOM: Major Command 주요사령부
MAPS: Modular Azimuth Position System 지형 자동측지장비
***MARFOR**: Marine Corps forces 해병대
MARS: Military Affiliates Radio System 군 가입 아마추어 무선체제
MAXI: Modular Architecture for Exchange of Intelligence 정보 상호교환장치
****MBA**: main battle area 주전투지역
MC: Military Committee 군사위원회

Movement Control 이동통제
**Marine Corps 해병대
MCC: Movement Control Center 이동관리대
MC&G: Mapping, Charting and Geodetic 지도, 도표제작 및 측지
MCI: Meal, Combat, Individual 전투식량
MCM: Mine Countermeasures 대 지뢰 방책
M-BM-S: Mobility, Countermobility, Survivability 기동, 대기동, 생존
MCRC: Master Control and Reporting Center 중앙방공관제소
M-Day: The day on which mobilization commences or is due to commence 동원개시일
MDC: Military District Command 관구 사령부
MDT: Mutual Defense Treaty 상호방위조약
****mech**: mechanized 기계화
Mech INF DIV: Mechanized Infantry Division 기계화보병사단
MEB (AMPHIB): Marin Expeditionary Brigade (Amphibious) 해병원정여단(상륙군)
MEB(MPS): Marine Expeditionary Brigade (Maritime Prepositioning Ships) 해병원정여단(해상사전배치선단)
MECL: Mission Essential Circuit List 임무필수회선목록
****MEDEVAC**: medical evacuation 의무환자 후송
MEDRES: Medium Resolution 중감도형
MEF: Marine Expeditionary Force 해병원정군
MERLIN: Magnetic Energy Reflect Linear Acceleration (자석 에너지 반사를 이용한 신형가속탄) 상보 공격용 박격포탄
METCOM: Meteorological Communications 기상통신망
****METL**: mission-essential task list 임무주요과제목록
METSAT: Meteorological Satellite troops, and time available 임무, 적, 지형 부대 및 가용시간
MEU: Marin Expeditionary Unit 해병원정부대
****mg**: machinegun 기관총
MGB: Medium Girder Bridge 간편조립교
MGSM: Multi-Sensor Ground Station Module 영상신호 종합 수신장비
****MHE**: materials-handling equipment 물자 취급 장비
****MHz**: Megahertz 메가헤르츠 (주파수 단위)
****MI**: Military Intelligence 군사정보
****MIA**: missing in cation 실종(자)
MIC: Military Intelligence Center 군 정보센터
****MICLIC**: mine-clearing line charge 지뢰제거 선형장약
****MIJI**: Meaconing, Intrusion, Jamming, or Interference 미지 (항법방해, 전자침투, 전자방해 및 전파간섭)
MILAN: Missile Infantry Light Anti-Armor 보병 경대전차 유도무기
****MILPO**: military personnel office(s) 군 인사처
****min**: minute(s) 최소의, 분
****MIRS**: multiple-launch rocket system 다연장 로케트 체제
MIRV: Multiple Independently

Targetable Reentry 다탄두핵미사일
MISDN: Military Integrated Services Digital Network 군 종합정보 통신망
MIW: Mine warfare 지뢰전
MLC: Martial Law Command 계엄사령부
MLR: Multi Launch Rocker 다연장 로케트(방사포)
MLRS: MLR System 다연장로케트체제
*__MMC__: materiel management center 물자관리 본부
MND: Ministry of National Defense 한국 국방부
MOA: Memorandum of Agreement 합의각서
MOC: Ministry of Communications 한국 체신부
Ministry of Construction 한국 건설부
MOG: Maximum-ON-Ground 최대 지상계류공간 (비행장)
**__MOGAS__: motor gasoline 휘발유
**__MOPMS__: modular-packed mine system 상자 살포식 지뢰
**__MOPP__: mission-oriented protective posture 임무형 보호태세
MOREX: Mobilization Exercise 동원훈련
**__MOS__: Military Occupational Specialty 군사주특기
**__MOUT__: Military Operations in Urban Terrain 시가지 군사작전
**__MP__: Military Police 헌병
MPAF: Ministry of the Peoples Armed Forces 인민무력부
**__mph__: miles per hour 시간당 마일 수
MPM: Medical Planning Module 의무기획작성 기본모형
MPS: Maritime Prepositioning Ships 해상사전배치선단
MR: Mobilization Reserve 해상사전배치선단
MR: Mobilization Reserve 동원예비군
MRD: Mobilization Reserve Launcher 동원사단
**__MRE__: Meal, Ready to Eat 전투식량
MRL: Multiple Rocket Launcher 다련장 로켓트포, 방사포
MRV: Multiple Reentry Vehicle 다탄두 미사일
MSCG: Medical Supply Coordinating Group 의무보급협조단
**__MSG__: master sergeant 상사
**__MSR__: Main Supply Route 주보급로
MTMC-TTC: Military Traffic Management Command - Transportation Terminal Command Far East 미 교통관리사령부 예하 극동수송단말 사령부
**__MTOE__: modified table of organization and equipment 수정 편제 장비표
MTT: Mobile Training Team 이동 훈련팀
MTZ: Motorized 차량화
MWD: Military Watch Dog 군견

N

N/A: Not Applicable 해당 무
NAD: Naval Ammunition Depot 해군 탄약창
NAI: Named Areas of Interest 관심표적 지역
NAR: No Answer (Action) Required 응신(조치)불필요
**__NATO__: North Atlantic Treaty Organization 북대서양 조약기구

*NAVFOR: Naval forces 해군

NAVOCEANCOM: Naval Oceanographic Command 해군 해양사령부

NAVSPACECOM: Naval Space Command 해군 우주사령부

NACSTAR: Navigation Satellite Tracking and Ranging 범세계 위치식별체계의 일종 (군용)

****NBC**: Nuclear, Biological and Chemical 화생방

NBCC: Nuclear, Biological and Chemical Center 화생방 통제소

N/C: No Change 변동사항 없음

***NCA**: National Command Authority 국가통수기구

NCAA: Non-nuclear Consumable Annual Analysis 제래식탄약 연간소모 분석회의

NCC: NAcal Component Command 해군구성군사령부

NCMA: National Command and Military Authority 국가통수 및 군사지휘기구

****NCO**: Noncommissioned Officer 부사관

****NCOIC**: noncommissioned officer in charge 책임 부사관

NCR: Naval Construction Regiment 미 해군 건설연대

****NCS**: net control station 무선 통신망 통제소

NDS: NUDET Detection System 핵폭발 탐지체제

NEDN: Naval Environmental Data 국가비상회의

***NEO**: Noncombatant Evacuation Operations 비전투원 후송작전

****NFA**: No Fire Area 사격금지구역

NFL: No Fire Line 사격금지선

****NGF**: naval gun fire 함포사격

****NGLO**: naval gunfire liaison officer 함포사격 연락장교

NIS: National Intelligence service 국가정보 통감

NK: Norht Korea(n) 북한

NKIL: North Korean Installation List 북한 시설목록

NKIS: Northern Limit Line 북방한계선

nm: Nautical Miles 해리

NMCD: Naval Mobile Construction Center 해군 기동건설본부

NOAD: Naval Ordnance Ammunition Depot 해군 병기탄약창

NOC: Naval Oceanographic Command 해군 해양사령부
Naval Operations Command 해군 작전사령부

NOD: Night Observation Device 야간관측장비

****NOE**: nap-of-the-erath 지형추적비행

NRT: Near-Real-Time 근 실시간

NSA: National Security Agency 국가 안정보장국

NSFS: Naval Surface Fire Support 사상함 화력지원

NSL: Nuclear Safety Line 핵피해안전선

***NSW**: Naval Special Warfare 해군 특수전

****NUC**: nuclear 핵

NUDET: Nuclear Detonation 핵폭발

****NVG**: night vision goggles 야시 장비(야간 투시경)

NWI: Northwest Islands 서북도서

O

OAC: Officers' Advanced Course 고등군사반

ORDR: Origination Agency's Determination Required 기안부서 결정

OASDPA: Office of the Assistant Secretary of Defense for Public Affairs 미 국방성 공보담당차관보실

****OB**: Order of Battle 전투서열

OBC: Officers' Basic Course 초등군사반

OCA: Offensive Counter-air 공세제공

***OCONUS**: outside the continental United states 미본토 이외의 지역

O-Kay : Organization Day 편성일, 부대 창설일

****OEG**: operational exposure guide 작전 노출 지침

****off**: officer 장교

****OIC**: officer in charge (of) 책임장교, 담당장교

OJT: On the Job Training 실무교육

OMG: Operation Maneuver Group 작전 기동단

****OP**: observation post 관측소

****POCON**: Operational Control 작전통제(권)

OPKS: Offshore Petroleum Discharge System 해안유류 분배작업

****OPLAN**: operation plan 작전계획

OPMG: Office of Provost Marshal General 헌병감실

****OPORD**: Operations Order 작전명령

****OPSEC**: Operations security 작전보안

OR: Operation, Research 운영분석

****ord**: ordnance 병기

OSCAR CP: Operations support Center and Rear Command Post 오스카 지휘소

OWR: Obligated War Reserve 필수전쟁예비물자

P, Q

PACDIGS: Pacific Digital Graphic System 태평양 디지털 화상체제

PACMEDS: Pacific Meteorological Data System 태평양 기상데이터 체제

PA: Public Affairs 공보

****PAO**: Public Affairs Office(r) 공보장교, 공보실(장)

PAR: Processing Analysis Reporting 분석, 처리, 보고

PARADROP: Parachute Drops 낙하산투하

PARPRO: Peacetime Application of Reconnaissance Program 평시 항공정찰계획

PASS-K: Pacific Command Automated Data Processing Server Site- Korea 미 태평양사 한국 자동화 처리소

PAT: Public Affairs Team 공보반

PATRIOT: Phased Array Tracking To Intercept of Target 중고고도 지대공 미사일

PB: Panel Bridge 조립교

PCM: Pulse Code Modulation 펄스코드 변조

PDE Cycle: Planning Decision and Execution Cycle 계획 결심 및 시행주기

PDM: Pursuit Deterrent Mine 휴대용 투척지뢰

PDS: Processing Display System 처리 시현체제

**personnel daily summary 일일 병력현황

PDSK: Pipeline Distribution system Korea 유류분배기구

****pers**: personnel 인원, 요원

PFT: Portable Flame Thrower 휴대용 화염방사기

PHIB: Amphibious 상륙

PHIBOPS: Amphibious Operations 상륙작전

PHOTINT: Photo Intelligence 사진정보

PIE: Peninsula Intelligence Estimate 한반도 정보판단

PIO: Public Information Officer 공보장교

**PIR: Priority Intelligence Requirement 우선정보요구

**PL: Phase Line 통제선

PLA: Plain Language Address 평문 주소, 평어 수신처

**PLD: Probable Line of Deployment 예상전개선

**PLL: prescribed load list 규정 휴대량 목록

**plt: platoon 소대

**PM: preventive maintenance checks and services 예방정비 점검 및 지원

POC: Point of Contact 담당자, 연락처

*POD: Ports of Debarkation 양륙항구

*POE: port of embarkation 탑재항, 적재항

**POL: Petroleum, Oils, Lubricants 가솔린, 석유, 오일, 구리스 등 각종 유류

POW: Prisoner of War 전쟁포로

PPDB: Point Positioning Date Base Program 측정선청 프로그램

PPWR: Prepositioned War Reserves 사전비축 전쟁예비

PRC: Peoples Republic of China 중공

Per-ITP: Prepositional Integrated Tasking Order 기계획 통합임무명령서

PREPFIGE: Preparatory Fire 공격준비사격, 예비사격

PSER: Personnel Status and Effectiveness Report 병력현황보고

PSG: Pershing Missile 퍼싱미사일

PSN: Packet Switching Node 통신망 단위 교환 노드

*PSRC: presidential selected reserve call-up

**PSYOP: Psychological Operations 심리전

PTC: Pacific Topographic Center 태평양 지형본부

Pub: Publication 발간

P-W: Pyongyang-Wonsan 평양-원산

**PW: Prisoner of War 전쟁포로

PWE: PW Enclosure 포로수용소

PWRM: Prepositioned War Reserve Material 사전배치 전쟁예비물자

PWRS: Prepositioned War Reserve Stock 사전비축 전쟁예비 재고

**PZ: pickup zone 인수 (회수) 지점

PZF-3: Panzerfaust-3 (전차잡는귀신) 경대전차무기

QRF: Quick Reaction Force 기동타격대, 신속조치군

R

RA: Regular Army 정규전

**RAANS: remote antiarmor mine system 살포식 대전차 지뢰 체제

RADCOM: Radio Communications 무선통신

RAG: Regimental Artillery Group 연대포병군

RAOC: Rear Area Operations Center 후방지역 작전본부

**RAP: rear area protection 후방지역 방호

PAS: Rear Area Security 후방지역 경계

RASIT: Radar Surveillance Intercept Terrain 지상감시 레이더

**RATT: radio teletypewriter 무선전신 타자수
RB: Reconnaissance Bureau 정찰국
RB: Ribbon Bridge 리본 부교
RBOC: Rapid-blooming Offboard Chaff 급속 체포 살포기
***RC**: reserve components 예비 구성군
RCC: Rescue Coordination Center 구조협조본부
RCU: Remote Control Unit 원격조정장치
RDF: Radio Direction Finding 무선 방향 탐지기
RDT&E: Research, Development, Test and Evaluation 연구, 개발 시험 및 평가
RECEXREP: Reconnaissance Exploitation Report 정찰판독 보고
****recon**: reconnaissance 수색
Regt: Regiment 연대
****REMAB**: remote marshalling base 원거리 작전 출동기지
****REMS**: remotely employed sensor 원격운용 감응장치
****repl**: replacement 보충(품, 병)
****req**: required 소요
****RF**: radio frequency 무선 주파수
****RFA**: restrictive fire area 사격 제한지역
****RFI**: radio frequency interference 무선 주파수 방해
****RFL**: Restricted Fire Line 사격제한선
RELROK: Releasable to Republic of Korea (i, e, Koreans) 한국측 열람가
RIF: Reduction in Force 병력감축
**reconnaissance in force 위력수색
RIL: Reconnaissance IL 정찰용 IL기
***RISTA**: reconnaissance, intelligence, surveillance, and target acquisition 정찰, 정보, 감시 및 표적획득
RMIG: Recconaissance MIG 정찰용 MIG기
RMTU: Reserve Mobilization Training Unit 예비군동원훈련부대
RO: Round-out 배속 (편제증원)
****ROE**: Rules of Engagement 교전규칙
ROK: Republic of Korea 대한민국
ROKA: Republic of Korea Army 한국 육군
ROKAF: Republic of Korea Fir Force 한국 공군
ROKAMC: Republic of Korea Army Mapping Center 한국 육군지도창
ROKG: Republic of Korea Government 한국 정부
ROKJCS: Republic of Korea Joint Chiefs of Staff 한국 합참
ROKMC: Republic of Korea Marine Corps 한국 해병대
ROK MND: Republic of Korea Ministry of National Defense 한국 국방부
ROK MOC: Republic of Korea Ministry of Communications 한국 체신부
ROKN: Republic of Korea Navy 한국 해군
ROKSF: ROK Special Force 한국특전부대 (육군)
ROKUS: Republic of Korea/United states 한미
ROTC: Reserve Officers Training Corps 예비역장교 훈련단
****RP**: release point (ground traffic) 분진점
***RPV**: remotely piloted vehicle 원격조정 무인 항공기
RRS: Remote Receive Server 전방 근접 영상수신 장비
R&S: Reconnaissance and Surveillance 정찰 및 감시

**RS: radiation status 방사능 현황
RS&O: Reception, Staging and Onward Movement
수용, 대기 및 전방이동
RSOP: Reconnaissance, Selection, and Occupation of Position 진지정찰, 선정 및 점령
**RSR: Required Supply Rate 소요 보급률
RSSC-K: Regional Signal Intelligence Support Center - Korea
주한 지역 신호정보 지원본부
RTC: Replacement Training Center
신병훈련소
**rte: route 경로, 축선, 접근로
RWCM: Regional Wartime Construction Manager
전시 후방지역 복구업체
**RWI: radio wire integration 유무선 통합망

S

SA: Security Assistance 안보지원
Selective Availability 선택적 가용성
SAC: Strategic Air Command 전략공군사령부
*SACEUR: Supreme Allied Commander, Europ
유럽 연합군 최고 사령관
SADARM: Sense and Destroy Armor Munition
감응식 대전차탄
SAEDA: Subversion and Espionage Directed Against the Army
미육군에 대한 전복 및 간첩행위
SAG: Surface Action Group
해상전투단
SAI: Significant Area of Interest
주요관심분야
SALS-K: Signal Ammuntition Logistics System - Korea 한미 단일 탄약지원체제
SALT: Strategic Arms Limitation Talks
전략무기제한협상
**SAM: Surface to Air Missile
지대공 유도탄
SAMLET: nk shore launched anti-ship missile
대함 해안유도탄
**SAMS: Standard Army Maintenance System
표준육군정비체제
SAR: Synthetic Aperture Radar
합성 구경 레이더
**SARTF: Search and Rescue Task Force
탐색 및 구조대
**SATCOM: Satellite Communications
위성통신
*SATP: Security Assistance Training Program
경계지원 훈련계획
SATRAN: Satellite Reconnaissance Advance Notice 위성정찰 사전경고
SB: Sniper brigade 저격여단
SB-Y: Switch Board-Y 교환기 - Y
SC: Search and Rescue Coordinator
탐색 및 구조협조관
SCATANA: security Control of Air Traffic and Navigational Aids
항공교통 및 항법시설의 보안통제계획
SCO: Security Control Officer
보안통제장교 (한국)
SCTY CLNC: Security Clearance
비밀취급인가
SCUBA: Selfcontained Underwater

Breathing Apparatus 잠수기구 (스쿠버)
SDC: Student Defense Corps
학도호국단
SDR: Source Directed Requirements
대상자 요구 첩보
****SEAD**: Suppression of Enemy Air Defenses
적 방공제압
***SEAL**: sea-air-land 해상, 공중 및 지상
SECON: Security Condition 보안태세
****SEE**: small emplacement excavator
소형 다목적 굴착기
SERE: Survival, Evasion, Resistance, and Escape
생존, 도피, 저항 및 탈출
SEWS: Satellite Early Warning System
위성조기경보체제
****SF**: Special Forces 특전부대
SFAF: Standard Frequency Action Format
표준 주파수 요청양식
SF Bde: Special Forces Brigade, Airborne
공수특전여단
SFG: Special Forces Group 특전단
SFGA: Special Forces Group, Airborne
공수특전단
SFOB: Special Forces Operational Base
특수작전기지
SHF: Super High Frequency 마이크로파
SHORAD: Short Range Air Defense
단거리 방공무기
SICARO: Secretary-in-Charge of Anti-ROK Operations 대남사업 담당비서
****SIDPERS**: Standard Installation / Division Personnel System 컴퓨터 시스템을 이용한 인사관리 제도의 일종
SI FAX: Special Intelligence Facsimile
특수정보 전송장비
****SIGINT**: Signal Intelligence 신호정보
SIPE: Soldier Integrated Protective Ensemble
통합방호 군장
****STMAP**: Situation Map 상황도
****SITREP**: Situation Report 상황보고
****SJA**: Staff Judge Advocate 법무참모
SLAR: Side Looking Airborne Radar
측방감시 레이더
SLGR: Small Lightweight Global Positioning System Receiver 소령 경량 GPS 수신기
SLOC: Sea Lines of Communication 해상병참선
SMA: Sergeant Major of the Army 육군 주임원사
SMC: Search and Rescue Mission Coordinator
탐색 및 구조임무협조관
SMO: Senior Meteorological and Oceanographic Officer 선임 기상 및 해양 담당관
SN: Service Number 군번
SO: Special Operations 특수작전
****SOC**: special operations command
특수작전 사령부
Security Control Officer 보안 통제장교
SOC-K: Special Operations Command - Korea
주한 특수작전사령부
SOCOM: Special Operation Command
특수작전사령부
****SOF**: Special Operation Forces
특수작전부대
SOFA: Status of Forces Agreement

한미 행정협정

**SOI: signal operation instructions 통신운용 지시

**SOP: Standing Operation Procedures 내규, 예규

SOSUS: Sound Surveillance System 음향 탐색체계

SOTA: SIGINT Operational Tasking Authority 신호정보 운용기구

**sp: start point 출발점

SPACC: Space Defense Operations Center 미 우주방어작전본부

SPCMCA: Special Court-Martial Convening Authority 특별군법회의 소집기구

SPITL: Single Prioritized Integrated Target List 단일 우선순의 통합표적 목록

SPOD: Sea Port of Debarkation 양륙항구

**spt: support 지원

**sqd: squad 분대, 반

Sqdn: Squadron 대대 (공군)

S-R: Secret - Republic of Korea / United States 한미 2급 비밀

SROKA: Second Republic of Korea Army 한국육군 제2군

SSAN: Social Security Account Number 주민등록번호

**SSB: Single side band 일변 주파수

SSI: Standing Signals Instructions 통신 준칙

SSP: Signal Intelligence Support Plan 신호정보 지원계획

S/T: Short Ton (2,000 pounds) 숏톤 (2,000 파운드)

STAMIDS: Standoff minefield Detection System 원거리 지뢰탐지 체계

**STANAG: Standardization Agreement 표준화 협정

**STRIKWARN: strike warning 타격경고

STW: Strike Warfare 타격전

subor: subordinate 예하부대

SUPIR: Supplemental Photographic Interpretation Report 추가 사진해석보고

SWC: Special Warfare Group 특전단

SWO: Staff Weather Officer 가상참모장교

T

TA: Table of Allowance 장배 배당표
Telephone Assembly 전화 조립체

TAA: Tactical Assembly Area 전술집결지역

**tac: tactical 전술적인

**TAC: Tactical Air Command 전술공군사령부
Tactical Air Coordination 전술항공 협조단

**TACAIR: Tactical Air 전술항공

**TACAN: tactical air navigation 전술공군 항법장치

**TACC: tactical air control center 전술항공 통제본부

TACCIMS: Theater Automated Command and Control Information Management System 전구 자동화 지휘, 통제, 정보관리체제

TACD: Tactical Deception 전술기만

TACE: Tactical Control and Analysis Element 전술통제 및 분석반

TACELINT: Tactical Electronic Intelligence
전술전자정보
*__TACON__: tactical control
전술적 통제
**__TACP__: tactical air control party
전술 보고서
**__TACSAT__: tactical satellite communi cation
전술위성 통신
TAI: Targeted Areas of Interest
중요표적지역
TANGO CP: Theater Air/Naval/Ground Operations Command Post 탱고지휘소
**__TAR__: tactical air reconnaissance
전술항공 정찰
TARS: Tactical Air Reconnaissance and Surveillance 전술항공정찰 및 감시
TAS: Tactical Air Support 전술항공지원
**__TASE__: Tactical Air Support Command
전술항공 지원반
*__TAV__: total asset visibility
**__TBA__: to be activated 추후 창설
**__TBD__: To Be Determined 추후결정
TBM: Tactical Ballistic Missile 전술 탄도탄
TBMW: Tactical Ballistic Missile Warning
전술탄도탄 경보
TBP: To Be Prepared/To Be Published
추후준비/추후발간
TD&A: Table of Distribution and Allowance
분배 및 할당표
TDT: Theater Display Terminal 전구시현단말기
TERS: Tactical Event Reporting System
전술사태 보고체제
*__TEWT__: tactical exercise without troops
전술 도상연습
**__TF__: Task Force 기동부대, 특수임무부대
TFC: Tactical fusion Center 전술융합센터
TFS: Tactical Fighter Squadron
전투비행대대
**__TFW__:(USAF) tactical fighter wing
(미공군) 전술전투 비행단
TG: Tear Gas 최루탄
**__tgt__: target 표적, 목표
TGW: Terminally Guided Warhead
종말 유도탄
**__TI__: technical inspection 기술 검사
TKP: Trans-Korea Pipeline 한국종단송유관
**__tm__: team 팀, 반
TM: Technical Manual 기술교범
Tm: Team 팀, 반
TMECS: Theater Mission Essential Communications System
전구임무 필수통신체제
**__TOC__: Tactical Operations Center
전술작전본부
TOD: Thermal Observation Device
열상관측장비
**__TOE__: Tables of Organization and Equipment 편성 및 장비표
TOPO: Topographic 지형
**__TOT__: Time On Target
일제사격, 티오티사격
*__TOW__: tube-launched, optically tracked, wire-guided 대전차 미사일, 토우
*__TPFDD__: Time-Phased Force and Deployment List

시차별 부대전개목록
TPT: Tactical Petroleum Terminal
전술 유류보급대
TRANSIT: Navy Navigation satellite System
미 해군 항법위성체제
TRE: Tactical Reconnaissance Element
전술정찰반
TRI-TAC: Tri-Services Tactical Communi cations
미 합동전술통신체제
TROKA: Third Republic of Korea Army
한국육군 제3군
TRT: Three Revolutions Teams 3대혁명 소조
TS: Team Spirit 팀스피리트
TSD: Taesong-dong 대성당
**TSOP: tactical standing operating procedure
전술 예규
TUGB: Training Unit Guidance Bureau
교대지도국

U

*UAV: unmanned aerial vehicle 무인 항공기
UBL: Unit Basic Load 기본휴대량
UDT: Underwater Demolition Team 수중폭파선
UFO: Ultra High Frequency Follow-on
UHF 후속통신 (차세대통신)
**UH: utility helicopter 다용도 헬기
**UHF: Ultra High Frequency 극초단파
*UK: United Kingdom 대영제국
UMN: Underwater Mine 수중기뢰
*UN: United Nations 유엔
*UNAAF: Unified Action Armed Forces
통합 작전군
UNC: United Natios Command 유엔군사령부
UNCSF: United nations Command Security Forces 유엔사 경비대
UNCSF-JSA: United Nations Command Security Forces - Joint Security Area
유엔사 공동경비구역 경비대
UNITREP: Unit Readiness Program
부대 준비태세 계획
U-R: Unclassified - Republic of Korea/United States 한미 평문
*US: United States 미국
USA: United States Army 미 육군
USAF: United States Sir Force 미 공군
USAMIIA: United States Army Medical Intelligence Information Agency
미 육군 의료정보첩보국
USARJ: United States Army Japan 주일미육군
USARPAC: United States Army Pacific
미 태평양육군
USARSPACE: United States Army Space Command 미 육군 우주사령부
*USCENTCOM: United States Central Space Command 미 육군 우주사령부
*USCENTCOM: United States Central Command 미 중부 사령부
*USCINCCENT: command-in-chief, United States 미군 사령관
*USCINCPAC: commander-in-chief, United States Pacific Command 미 태평양 사령관
*USCINCSO: commander-in-chief, United States Southern Command
미 남부 사령관

***USEUCOM**: United States European Command 미 유럽 사령부

USFJ: United States Forces Japan 주일 미군

***USFK**: United States Forces Korea 주한미군

USIA: United States Information Agency
미 정보기관

USIS: United States Information Service
미 첩보국

USJCS: United States Joint Chiefs of Staff
미 합동참모본부, 미 합참

USMARFORK: United States Marine Forces Korea 주한 미 해병대

***USMC**: United States Marine Corps 미 해병대

USMTF: United States Message Text Format
미 표준전문양식

****USN**: United States Navy 미 해군

USNAVFOR: United states Naval Forces
미 해군부대

USPACAF: United States Pacific Air Force
미 태평양공군

USPACFLT: United States Pacific Fleet
미 태명양함대

USPACOM: United States Pacific Command
미 태평양사령부

***USSOCOM**: United States Special Operation Command 미 특수작전사령부

***USSOUTHCOM**: United States Southern Command 미 남부 사령부

***USSPACECOM**: United States Space Command 미 우주사령부

USSTRATCOM: United States Strategic Command 미 전략사령부

USTRANSCOM: United States Transportation Command 미 수송사령부

****UTM**: Universal Transverse Mercator
유티엠 좌표

UW: Unconventional Warfare

V W X Y Z

****VHF**: Very High Frequency 초단파

****vic**: in the vicinity of ---부근에

****VIP**: Very Important Person 귀빈

VOB: Voice Operation Branch 감청

VOQ: Visiting Officer's Quarters 방문장교숙소

****VT**: variable time 가변 시간

VTC: Video Tele-Conference 화상회의

****VTOL**: vertical take-off/landing 수직이착륙

WAC: Women's Army Corps 여군단

WAN: Wide Area Network 광역 통신망

WARM: Wartime Reserve Modes 전시예비모드

****WCS**: weapons control status 무기통제상태

WDT: Warning Display Terminal 경보시현단말기

WG: Weather Group 기상반

WGS: world Geodetic System 세계좌표체계

WHNS: Wartime Host Nation Support
전시 (주둔국) 지원

WIS: WWMCCS Information System
범세계 군사지휘통제체계 정보체계

WMR: Wartime Movements Requirements 전시 이동소요
****WO**: Warning Order 준비명령
****WP**: white phosphorus 백린 연막탄
*****wpn**: weapon 화학, 무기
WR: War Reserves 전쟁 예비
WROE: Wartime Rules of Engagement 전시교전 규칙
WRS: War Reserve Stock 전쟁예비 재고
WRSA: War reserve Stocks for Allies 동맹국을 위한 전쟁예비 재고
WS: Weather Squadron 공군 기상대대
WSA: Weapon System Analysis 무기체제분석
WSMC: Western Space and Missile Center 서부 우주미사일본부
WSR: Wartime Sustainability Review 전시 지속능력회의
WWMCCS: Worldwide Military Command and Control System 범세계 군사지휘통제체제
ZF: Zone of Fire 사격 (발사)지대
2ID: 2d Infantry Division (US) 미 제2사단

김정기

육군사관학교 졸업

군사영어반 수료

연세대학교 석사

경남대학교 박사

현 연성대학교 교수(국방군사학과장)

현 한국군사학회 이사

남기봉

육군사관학교 졸업

경남대 행정대학원 석사

육군 대령 예편

전 연성대학교 교수

천종웅

육군사관학교 졸업

고려대학교 석사

King's College London 석사

광운대학교 박사

Oxford대학교 방문연구원

현 연성대학교 겸임교수